AF798955

Law for Professionals

The *Law for Professionals* book series contains publications in all fields of the law aimed at professionals and practitioners. The book series is dedicated to publish high quality books that provide in-depth insights into the workings and the practical functioning of the law, international organizations or regional legal frameworks.

The authors contributing to the book series are experienced legal professionals and practitioners whose expertise combines theory and practice.

Javier Guillem Carrau

Landmark EU Geographical Indications Case Law

Key Decisions Shaping the Protection of Origin and Quality

Springer

Javier Guillem Carrau
EUIPO
Alicante, Spain

University of Valencia
Valencia, Spain

Regional Parliament of Valencia (Corts Valencianes)
Valencia, Spain

ISSN 2662-141X ISSN 2662-1428 (electronic)
Law for Professionals
ISBN 978-3-032-26100-7 ISBN 978-3-032-26101-4 (eBook)
https://doi.org/10.1007/978-3-032-26101-4

© The Editor(s) (if applicable) and The Author(s), under exclusive license to Springer Nature Switzerland AG 2026

This work is subject to copyright. All rights are solely and exclusively licensed by the Publisher, whether the whole or part of the material is concerned, specifically the rights of translation, reprinting, reuse of illustrations, recitation, broadcasting, reproduction on microfilms or in any other physical way, and transmission or information storage and retrieval, electronic adaptation, computer software, or by similar or dissimilar methodology now known or hereafter developed.
The use of general descriptive names, registered names, trademarks, service marks, etc. in this publication does not imply, even in the absence of a specific statement, that such names are exempt from the relevant protective laws and regulations and therefore free for general use.
The publisher, the authors and the editors are safe to assume that the advice and information in this book are believed to be true and accurate at the date of publication. Neither the publisher nor the authors or the editors give a warranty, expressed or implied, with respect to the material contained herein or for any errors or omissions that may have been made. The publisher remains neutral with regard to jurisdictional claims in published maps and institutional affiliations.

This Springer imprint is published by the registered company Springer Nature Switzerland AG
The registered company address is: Gewerbestrasse 11, 6330 Cham, Switzerland

If disposing of this product, please recycle the paper.

Foreword

It's not every day that you get the chance to read a book that so deeply explores a subject that has shaped the author's career. This book is exactly that—a thoughtful analysis of the intricate world of Geographical Indications (GIs) and the legal framework that supports them within the European Union (EU), written by a leading authority.

Having worked in the Commission Legal Service for almost thirty years, and having dealt for a few years with the issues covered in the book, I've seen firsthand the passion and depth of knowledge that Javier Guillem brings to this subject. Throughout his career, from his doctoral thesis to his subsequent work both at the European Commission's Legal Service, advising the relevant department and representing the Commission in landmark GIs cases before the European Court of Justice, and at the EUIPO, Javier Guillem has amassed an unparalleled insight into the world of GIs, from different angles. The main objective of the EU GIs scheme is to promote quality, variety, and value in the food chain. The EU's GIs system relies on the fact that qualities and characteristics are due to the specific geographical area of origin. One of the key challenges of the EU's GIs scheme is striking a balance between protecting the intellectual property rights of producers and ensuring that the system is fair to all.

It has been nearly five decades since the first EU regulation on GIs was enacted. In this half-century, a variety of pivotal events have contributed significantly to the consolidation and evolution of GIs in the EU. I would like to highlight two undercurrents in this process, which shed light on the essence of GIs. Firstly, their seamless integration into the field of intellectual property law, drawing inspiration from trademark law, yet transcending its boundaries to provide a broader scope of protection. Secondly, through case law and legislative reforms, GIs have undergone a process of gradual evolution, ultimately leading to their extension to encompass artisanal products, as a sign of the adaptive nature of this legal framework.

The importance of GIs has been widely acknowledged by scholars approaching the subject from various angles. However, a more comprehensive analysis was needed, one that would examine the European Court of Justice's case law in depth and locate it within the general parameters of the EU GIs scheme. This requires a very close look at the evolutionary character of the problems that have appeared along the road and how the Court's judgments and legislative reforms have shaped this area. By doing so, we can gain a deeper understanding of how EU GI law has developed until its current design under Regulation (EU) 2024/1143. This is what Javier Guillem does in this book with precision, rigour, and expertise, avoiding lengthy and unnecessary explanations.

This book is the culmination of years of research, practice, and firsthand experience. The book explores, chapter by chapter, all the different aspects of the EU's GIs scheme: its objective and how it is part of the Common Agricultural Policy, the conditions for registration, the scope of protection, its implementation and enforcement, labelling rules and its coexistence with other forms of intellectual property. It does so by focussing on landmark decisions that have shaped the protection of GIs, like *Feta*, *Rioja II*, and *Aceto Balsamico di Modena*, among many others. From the protection of the *terroir* to the coexistence of GIs with other intellectual property rights like trademarks, the author acknowledges that GIs do not live in a legal autarchy and explores the broader implications for intellectual property law. This complex legal framework and the different cases that have shaped its evolution are presented in a way that is comprehensive, accessible, and engaging.

The book now published by Javier Guillem meets two of the conditions of a good legal monograph: timeliness and usefulness. Whether you're an expert in the field or someone new to the subject, this work provides invaluable insights into the objectives, procedures, and challenges surrounding GIs. It offers a close look at how the protection of traditional know-how, regional quality, and sustainability is being balanced today within EU law.

It's a true pleasure to see this work come to life, and I'm excited for you, the reader, to explore the world of GIs with the guidance of someone who truly understands the complexities and the importance of this fascinating field. I shall stop here in my presentation of the book, leaving the reader the pleasure of discovering and appreciating its many merits. Congratulations!

Legal Service of the European Commission
Brussels, Belgium

Fernando Castillo de la Torre

Preface

Welcome to "Landmark EU Geographical Indications Case Law: Key Decisions Shaping the Protection of Origin and Quality". I am delighted to present this book on EU geographical indications (GIs) case law.

As an expert in the field of EU law and GIs, I have always been fascinated by the issues surrounding the protection of origin and quality of products. It all began with my training period in DG VI (now AGRI) of the European Commission and my doctoral thesis, which I defended at the University of Valencia's Faculty of Law. Then, I've had the privilege of working with my colleagues of the European Commission's Legal Service and acting before the Court of Justice on GIs cases. I've also had the opportunity to engage with experts at the European Union Intellectual Property Office (EUIPO). These experiences have given me a unique perspective on GIs, and I'm excited to share my insights with you.

So, what can you expect from this book? This book provides an in-depth analysis of the key decisions of the EU Courts that have had a significant impact on the protection of GIs.

I am grateful to Springer for providing me with the opportunity to publish this book and share my research with a wider audience. The team at Springer has been a pleasure to work with, and I appreciate their professionalism and expertise in bringing this book to publication.

As you read this book, I hope you will gain a deeper understanding of the law and practice of GIs in the EU. It is about a readable text on EU GIs law that allows the reader to understand how the system works and to have a clear and helpful overview of the EU jurisprudence. I hope you will find the analysis of the key decisions to be useful in your own work or research. And I hope you will join me in exploring the evolving world of EU GIs case law.

Gaining insight into the judgments related to GIs is essential to understanding how the complex puzzle of GI protection fits together within EU law. This system empowers producers to manage their respective GIs, enabling them to respond to growing societal demands for sustainable products. The scope of protection is carefully calibrated to balance the intellectual property rights of agrifood GIs with the coexistence of pre-existing trademarks. Notably, the protection of GIs extends beyond mere conceptual evocation, also encompassing cases involving figurative signs and even product shapes. A review of European Court of Justice and General

Court case law reveals a clear trend in favour of protecting GIs, reinforcing the importance of this intellectual property right.

This book is intended for anyone interested in EU law, GIs and their coexistence with trademarks. It is my hope that it will be a valuable resource for practitioners, policymakers, researchers, and students, and that it will contribute to a greater understanding and appreciation of the importance of protecting GIs in the EU.

Finally, I am honoured to have Fernando Castillo's foreword. It is a wonderful addition to this book, and I thank him for his generosity in writing it. I also would like to express my admiration for his work in the European Commission's Legal Service and for the example he has set for us all through his commitment to public service and the values of the European Union.

This book also is for my family, who've inspired me every step of the way.

Thank you again to Springer for this opportunity and thank you to my readers for their interest. I hope you enjoy it, and I look forward to hearing your thoughts and feedback.

Sant Joan d'Alacant, Spain Javier Guillem Carrau

Acknowledgement

The opinions set out in the article do not represent the views and opinions of the EUIPO, but solely the personal views and opinions of the author.

Competing Interests The author has no competing interests to declare that are relevant to the content of this manuscript.

Contents

Introduction

1

Abstract

This book examines the evolution, controversies, and procedural aspects of the EU's Geographical Indications (GIs) scheme, a cornerstone of the Common Agricultural Policy and the internal market. Through a consistent analysis of landmark case law—from *Feta* and *Prosciutto* to *Rioja II* and *Schwarzwälder Schinken*—it explores how GIs have become a solid component of the EU's Policies, shaping agriculture, trade, intellectual property, and sustainability agendas. The chapters scrutinise the legal foundations of GI protection, including the registration process, exclusivity of EU titles, and the role of Member States and the Commission in protecting them, while addressing challenges such as national court reviews, infringement measures, and the impact of the Digital Services Act on online enforcement. A critical focus is placed on the balance between origin, specificity, and the threat of genericity, with judgments like *Grana Padano* and *Aceto Balsamico di Modena*, and illustrating the backbone of the expansive protection against misuse, evocation, and misleading practices with judgments like *Scotch Whisky, Manchego, Morbier and Champanillo*. The book further analyses labelling rules, Union symbols, and the intersection of GIs with trade marks, drawing on cases such as *Nero Champagne. Champagner sorbet and Weingut* cases serve as example of the relevance of the enforcement of the GIs as any other intellectual property right. By fusing judicial interpretations, procedural hurdles, and enforcement mechanisms, this work provides professionals—whether legal practitioners, policymakers, or industry stakeholders—with a rigorous, case-law-driven guide to navigating the complexities of GI protection in the EU's dynamic regulatory landscape.

© The Author(s), under exclusive license to Springer Nature Switzerland AG 2026

J. Guillem Carrau, *Landmark EU Geographical Indications Case Law*, Law for Professionals, https://doi.org/10.1007/978-3-032-26101-4_1

At EU level, unitary GIs protection was traditionally provided for wines, spirit drinks, aromatised wines and for agricultural products and foodstuffs.[1] Nowadays, there is EU Law on GI protection in the fields of agricultural products and foodstuffs[2] and for craft and industrial products.[3]

The main objective of the so-called EU GIs scheme is to promote quality, variety, and value in the food chain, together with the preservation of traditional know-how in the sector and to stimulate diversification and employment in rural areas. Through this EU scheme, protected names for the covered products enjoy a far-reaching and unitary protection applicable on the whole territory of the EU, based on a single application.

This book provides an in-depth examination of the EU GIs scheme, a multifaceted system that has evolved over the years. In Chap. 2, we explore the EU's objectives for the GIs scheme, including its relationship with the Common Agriculture Policy and the Internal Market. The chapter sets the stage for the book by examining the controversies surrounding the scheme, highlighting cases such as *Feta, Prosciutto, Mozzarella, Rioja II* and *Schwarzwälder Schinken.* Likewise, *Commission v Council (Revised Lisbon Agreement)* case serves to demonstrate that GIs also constitutes a substantial component of the Common Commercial Policy and it shows a foothold to extend the scope of GI protection under the Geneva Act.

As we dig deeper into the EU GIs scheme, we examine the key components that underpin the system through analysis of the General Court and Court of Justice (CG/CJ) jurisprudence. In Chap. 3, the review of a jigsaw of case law as *Rioja II, Turrón de Jijona* or *Salame Felino* breaks down the essential components of the EU GIs including its place within the intellectual property rights framework, the requirement for EU registration, and the concept of exclusivity, as said in *Budějovický Budvar* case. EU GIs system's genuine character relies on the fact that qualities and characteristics are due to the specific geographical area of origin, as described in *Sekt and Weinbrand,* and the GIs intrinsic reputation, as described in *Bavaria Judgment* or in *Nero Champagne* Boards of Appeal's Decision. *Grana Biraghi* Judgment determines that GIs protection depends on the legislative protection and its historical development and not on definitions of the dictionaries or internet extracts. Within the set of components, the chapter also explores the protection of the *terroir* and the *savoir faire* of the product, associated

[1] Dr Javier Guillem Carrau, IP Specialist, EUIPO. Associated Professor of the Department of Commercial Law 'Manuel Broseta Pont', University of Valencia. Legal Adviser, Regional Parliament of Valencia Region (Corts Valencianes). The opinions set out in the book do not represent the views and opinions of the EUIPO, but solely the personal views and opinions of the author.

[2] Regulation (EU) 2024/1143 of the European Parliament and of the Council of 11 April 2024 on geographical indications for wine, spirit drinks and agricultural products, as well as traditional specialities guaranteed and optional quality terms for agricultural products, amending Regulations (EU) No 1308/2013, (EU) 2019/787 and (EU) 2019/1753 and repealing Regulation (EU) No 1151/2012. *OJ L, 2024/1143, 23.4.2024.*

[3] Regulation (EU) 2023/2411 of the European Parliament and of the Council of 18 October 2023 on the protection of geographical indications for craft and industrial products and amending Regulations (EU) 2017/1001 and (EU) 2019/1753. PE/31/2023/REV/1, *OJ L, 2023/2411, 27.10.2023.*

to its origin, that can benefit only the eligible producers, as declared in *Manchego* and how generic terms are not protected because the protection is reserved to the protected term as such, as clarified in *Aceto Balsamico di Modena*; last but not least, it focus on the concept of sustainability as a novelty for the EU GIs scheme. This examination attempts to draw conclusions on the extent of the GIs protection and to establish overall conclusions on the connection with the principles of, among others, veracity of origin, reputation, exclusivity and sustainability.

You will appreciate that the EU GIs scheme is not just a theoretical framework, but a practical system that requires careful implementation and enforcement. Chapter 4 covers GIs procedural aspects regarding registration, controls, and enforcement. Through a series of cases, including the *Port Charlotte, Halloumi cheese,* and *Comté* cases, this chapter examines the importance of registration as a condition for protection, the role of Member States in facilitating the reception of the application, the existence of a provisional protection, the two types of amendments of the GI' specifications and the particularities of the national opposition procedures. *Kühne and Jambon sec* cases showed that the general understanding should be that it is for the Member State to check the application; for the Commission to assess that the application contains the information required (i.e. standing orders) and to scrutinise it is not vitiated with errors; and for the National Courts to verify decisions taken by national authorities on minor amendments. Likewise, we make technical findings on the difficulties that the system faces when the national decision is reviewed by a national Court or when the Member State is called to adopt measures against conduct which infringe a PDO. Furthermore, it describes the Member States' responsibility on the verification and control systems in place to ensure the authenticity and quality of GI products and the new responsibilities for the judicial or administrative authorities of the Member States as regards the GIs enforcement and the online platforms under the EU Digital Services Act.

One of the key challenges of the EU GIs scheme is striking a balance between protecting the intellectual property rights of producers and ensuring that the system is fair to all. Chapter 5 answers questions related to the examination of GIs and the balance between origin and specificity, and the threat of genericity through a series of cases, including *Feta, Grana Padano, Edam Holland,* and *Aceto Balsamico di Modena*. By exploring the real scope of GI protection, it attempts to describe how the Court explained the contours of the prohibition of direct and indirect use, misuse, and misleading practices, as seen in cases such as *Bavaria, Chianti*, and *Scotch Whisky Association*. Furthermore, it seeks to provide an appropriate analysis of the criteria for assessing comparable products, as exemplified by the *Cognac (BNIC)* case, and the expansive protection against evocation, as demonstrated by *Manchego, Morbier* and *Champanillo*. Finally, it touches on the extension of GI protection under unfair competition law, reflecting on the *Salami Felino* case.

The EU GIs scheme also has important implications for labelling and packaging, as producers must comply with strict rules and regulations to ensure that their products are accurately labelled and presented according to the Food Labelling Regulation and according to our practice. In Chap. 6, GIs labelling rules are described as an expression of the exclusive use of the GI. In this part, the rules on the use of Union symbols on the packaging (labelling and advertising material) are examined with the prism of the Commission Guidelines on labelling practices for GIs and their incorporation to the legal text after its last reform. *Champagner* sorbet and *Weingut* cases serve to understand how the protection should work when a GI is used as an ingredient and to illustrate about some specificities of labelling and presentation in GI wines, respectively.

Finally, the EU's GIs scheme must coexist with other forms of intellectual property rights, such as trade marks, collective trade marks and certification trade marks. Chapter 7 is about the points of intersection and conflict of GIs and trade marks. It is an attempt to explain the coexistence throughout the key decisions rendered by the EUIPO Boards of Appeal on GIs, particularly in relation to the absolute ground of refusal under Article 7(1)(j) EUTMR. It also covers the role of GIs in inter partes proceedings, by analyzing a decisions like *Campo Verde, Toro de Piedra, and Chenine Brut Premium Cava Bottle, Colombiano Coffee House, Romeral de Jabugo Padrón Auténtico and Marqués de la Mancha.* Through these cases, and other preliminary rulings as *Manchego, Scotch Whisky, Morbier and Champanillo,* it sheds light on the scope and extension of GI protection in trade mark disputes, providing insights into the interactions between these two intellectual property rights. *Nero Champagne* Judgment confirmed that EU law does not prohibit a trade mark from containing a GI and it showed how to apply the so-called limitation theory. The analysis of the *Darjeeling tea* and *Halloumi* Judgments shows how the interplay with collective trade marks can weak the GI. This part ends with a description of the differences between GIs and the certification trade marks.

To sum up, this book provides a comprehensive examination of the EU's GIs scheme, from its objectives and context to its essential components, procedural aspects, protection and balance, labelling rules, and coexistence with trade marks. By exploring the pillars of the system, you can gain a deeper understanding of the challenges of implementing the EU's GIs scheme and appreciate the need for a balanced approach to protecting intellectual property rights.

Geographical Indications in the EU: A Multifaceted Policy Framework

2

Abstract

Over the years, the EU established the objectives of the quality schemes for GIs in the context of the common agriculture policy and the internal market. As it will be explained in this chapter, the set of goals of the EU GIs scheme has not been, of course, uncontroversial, like the Feta, Prosciutto and Mozzarella cases illustrate. Likewise, Rioja II and Schwarzwälder Schinken cases shaped the GIs position in the context of the free movement of goods. Then, Commission v Council (Revised Lisbon Agreement) case serves to demonstrate that GIs have become a substantial component of the Common Commercial Policy and provided a foothold to extend the scope of GI protection under the Geneva Act.

Agri-food and drink products whose names are protected by the European Union as "Geographical Indications" (GIs) represent a sales value of €74.76 billion. Over one fifth of this amount results from exports outside the European Union. A study collected and analysed economic data on Protected Designation of Origin (PDO), Protected Geographical Indications (PGI) and Traditional Specialities Guaranteed (TSGs) registered in the, at that moment, 28 Member States of the European Union, over the 2011–2017 period (3153 GIs and 54 TSGs).[1]

[1] The estimate of sales value of GIs/TSGs was EUR 77.1 billion in 2017 (EUR 74.8 billion excluding TSGs), wines accounting for 51% of this value (EUR 39 billion), agricultural products and foodstuffs for 35% (EUR 27 billion), spirit drinks for 13% (EUR 10 billion), and aromatised wine products for 0.1% (EUR 43 million). The total sales value covered by GIs/TSGs has grown by 42% since 2010 (37% excluding TSGs). Five Member States had a sales value of GI products over EUR 5 billion each: France, Italy, Germany, the UK and Spain. The value of whole exports (intra-EU trade and exports to non-EU countries) of GI/TSG products was estimated to represent EUR 32.10 billion, which accounted for 42% of the total sales value in 2017 (20% for intra-EU trade and 22% for exports to non-EU countries). Finally, the global value premium reached EUR 40 billion in 2017 and increased by 38% compared to 2010. The value premium rate moved

© The Author(s), under exclusive license to Springer Nature Switzerland AG 2026

J. Guillem Carrau, *Landmark EU Geographical Indications Case Law*, Law for Professionals, https://doi.org/10.1007/978-3-032-26101-4_2

Ultimately, the EU legislator has adopted Regulation (EU) 2024/1143 and repealed Regulation (EU) 1151/2012 with the objective of granting a uniform recognition and protection across the Union for the intellectual property rights related to names protected in the Union[2] and, in parallel, Regulation (EU) 2023/2411[3] ruled a new GI scheme for craft and industrial products.[4] In recital 18 Regulation (EU) 2024/1143, the legislator recalls that:

> … ensuring uniform recognition and protection throughout the Union for the intellectual property rights related to names protected in the Union is a priority that can be effectively achieved only at Union level. A unitary and exhaustive system of GIs therefore needs to be provided for in Union law. GIs are a collective right held by all eligible producers in a designated area willing to adhere to a product specification.

Since 1992, the EU GIs scheme contemplates two categories, PDO and PGI, with a close or inseparable correlation between a product and a specific territory, considering only the natural and human factors that explain the specificity of a product, without a prior or inherent evaluation of the objective quality of the protected product. It is evident that a product covered by a PDO or PGI complies with European law (commercial norms, health conditions, etc.) by virtue of the law itself.[5]

---

from 2.14 in 2010 to 2.07 in 2017. See EUROPEAN COMMISSION (2019) Study on economic value of EU quality schemes, geographical indications (GIs) and traditional specialities guaranteed (TSGs), Final report, accessible https://op.europa.eu/en/publication-detail/-/publication/a7281794-7ebe-11ea-aea8-01aa75ed71a1/language-en.

[2] Regulation (EU) 2024/1143 of the European Parliament and of the Council of 11 April 2024 on geographical indications for wine, spirit drinks and agricultural products, as well as traditional specialities guaranteed and optional quality terms for agricultural products, amending Regulations (EU) No 1308/2013, (EU) 2019/787 and (EU) 2019/1753 and repealing Regulation (EU) No 1151/2012, OJ L, 2024/1143, 23.4.2024.

[3] Regulation (EU) 2023/2411 of the European Parliament and of the Council of 18 October 2023 on the protection of geographical indications for craft and industrial products and amending Regulations (EU) 2017/1001 and (EU) 2019/1753. PE/31/2023/REV/1, *OJ L, 2023/2411, 27.10.2023.*

[4] What's good for the goose is good for the gander, goes the saying. EU craft GI system provides a high level of protection for registered names, the same than the EU legislation on food GIs: See Guillem Carrau (2023), p. 728.

[5] González Sacedón and Vassilakis (1998), p. 37.

Over the years,[6] the Union has established the objectives of the quality schemes for products with identifiable specific characteristics which cover GIs[7] for wine, spirit drinks and agricultural products, including foodstuffs, as well as traditional specialities guaranteed and optional quality terms for agricultural products, including foodstuffs in the context of the Common Agriculture Policy (CAP) and the internal market.[8] Likewise, GIs have been a pivotal component of the Common Commercial Policy as can be gathered by the role they play in the negotiations of international agreements by the European Commission.

In this context, the set of policy goals of the EU GIs scheme has not been, of course, uncontroversial, as the *Feta, Prosciutto, Mozzarella, Council vs Commission* cases illustrate.[9]

2.1 A Sound European Policy in the Context of the Internal Market

In 1990s, the establishment of an EU framework and EU-wide protection seemed appropriated within the context of realising an internal market.[10] On the top of that, in the agricultural sector, the preference of the EU legislator was to recognise a connexion between the origin and the quality and the quality that can show different meanings.[11] Likewise, the legislator had not a lot of options as the legal basis of the

[6] As regards the evolution of the EU GIs scheme, the 2012 reform did not bring about a substantial change in the definition of the two categories that serve as the cornerstone of the EU's agri-food quality system. However, it is necessary to highlight a difference that, despite seeming minor, is of great importance due to its practical implications. Where the previous regulation (CE No. 510/2006) mentioned "production, processing, and preparation", the new regulation (CE 1151/2012) only mentions "production". Additionally, as an exception to the general rule for DOP, certain raw materials (live animals, meat, and milk) may be identified as DOP when the production area is delimited, specific conditions for their production exist, and control measures are in place to guarantee this (Article 5.3, final part). The system was reformed so that the DOP category remains the pinnacle, as the product will evoke a place and its quality will be fundamentally or exclusively due to the geographical environment, including human and geographical factors, with all production phases taking place in that origin. The IGP category was reserved for a lower level, for products that evoke a place and whose determined quality, reputation, or other characteristic can be essentially attributed to their geographical origin, with at least one of the production phases taking place in that origin.

[7] For ease of reading, we will use the term GI to refer to both categories, PDO and PGI, wherever possible.

[8] Among other, see Favret (2000), Mattera (1991) and Swann (1992) for the internal market whole picture; and as regards GIs' perspective, see Calboli (2017), p. 3; Gangjee (2012), p. 265; Kur and Cocks (2007), p. 999.

[9] C-108/01, Consorzio de Prosciutto di Parma and Salumificio S. Rita, EU:C:2003:296; C-289/96, C-293/96 and C-299/96, Feta ECLI:EU:C:1999:14, ECLI:EU:C:2005:636; C-569/18, Caseificio Cirigliana and Others, ECLI:EU:C:2019:873C-389/15, Council vs Commission, ECLI:EU:C:2017:798.

[10] Brouwer (1991), p. 615.

[11] Audier (2003).

EU GIs scheme referred only to agricultural products and it would have been not feasible to extend the scope of the legal instrument to industrial goods.[12]

Of course, one needs to consider that GIs are a "special animal" in terms of their substance and history; they cannot simply be put on equal footing with any other IP right. At their root lies the concept of *terroir* giving certain products their unique characteristics and quality due to the natural surroundings. This notion gave GIs their special place in the universe of IP law and justified their broad protection.[13]

This is also valid for the digital markets, which are regulated under the EU Digital Services Act,[14] and where the GIs will have a specific treatment by the legislator in the context of the enforcement since 2024, as described in Chap. 4.

2.1.1 The Controversial Objectives of the GIs System: *Prosciutto di Parma* and *Commission v Denmark (PDO Feta)* Cases

In *Prosciutto di Parma*,[15] the House of Lords referred to the Court for a preliminary ruling a question on the interpretation of the GIs system in the context of the internal market. In particular, the question referred to the specifications of the PDO 'Prosciutto di Parma' (Regulation No 1107/96). These specifications expressly mentioned the requirement of slicing and packaging the product in the region of production for ham marketed in slices and referred to the Law of 13 February 1990 and the Decree of 15 February 1993 as requirements laid down by national provisions. The specifications, which were defined as such to grant the protection of the quality and genuine character of the ham, constituted a measure having equivalent effect to a quantitative restriction on exports within the meaning of Article 35 TFEU but may be regarded as justified and proportional, and hence compatible with that provision. The Judgment demystified the complexity of the system which creates an exemption on the applicability of the free movements of goods to grant the genuine character of the goods produced according to the specifications and its circulation in the markets.

A second noteworthy case is the *Feta cheese* case. As regards the name '*Feta*', it was registered as a PDO in 1996, but the Court of Justice annulled its registration because the Commission did not take due account of all the factors required by the applicable law. Following that judgment, *Feta* was withdrawn from the EU GIs Register.[16] Later, by letter of 15 October 1999, the Commission sent the Member States a questionnaire

---

[12] Beier and Knaak (1994), p. 38; Oliver (1996), p. 299.

[13] Kur (2023), p. 87.

[14] Regulation (EU) 2022/2065 of the European Parliament and of the Council of 19 October 2022 on a Single Market For Digital Services and amending Directive 2000/31/EC, OJ L 277, 27.10.2022.

[15] Op. cit 9.

[16] The Commission adopted Commission Regulation (EC) No 1070/1999 of 25 May 1999 amending the Annex to Regulation (EC) No 1107/96 (OJ 1999 L 130, p. 18), which deleted the name 'feta' from the Register of PDOs and PGIs and from the Annex to Regulation No 1107/96.

on the manufacture and consumption of cheeses known as 'feta' and on how well known that name was amongst consumers in each of the States. The information received in response to that questionnaire was presented to the scientific committee, which gave its opinion on 24 April 2001. In that opinion, the committee concluded unanimously that the name 'feta' was not generic in nature. On 14 October 2002, the Commission added *Feta* to the Register.[17] The registration of *Feta* was contested but the actions were dismissed, and the Court concluded that the statement of reasons contained in the contested regulation was sufficient for a finding that the name 'feta' was not generic.[18] Therefore, in 2002, the name 'Feta' was registered as a PDO. Since then, that name can be used only for cheese that originates in the defined geographical area in Greece and conforms to the applicable product specification.

Feta is indeed relevant from the point of view of the Internal market as showed the case *Commission v Denmark (PDO Feta).*[19] The main contribution of this case consisted of the fact that, for the Court, the use of a GI happens even when the product is in transit for being exported to third countries. The Court of Justice noted in the first place that the use of a registered name to designate products not covered by the registration which are produced in the EU and intended for export to third countries is not excluded from the extent of the protection of *Feta* PDO.

In this case, the Commission, supported by Greece and Cyprus, claimed that Denmark breached its obligations under Regulation (EU) No 1151/2012 by failing to prevent or stop the use of the designation 'Feta' on cheese produced in Denmark and intended for export to third countries. The use of a PDO or PGI to designate a product produced in the territory of the EU which does not comply with the applicable product specification impairs, within the EU, the intellectual property right constituted by that PDO or PGI, even if that product is intended for export to third countries. In the third place, regarding the objectives pursued by the applicable Regulation No 1151/2012, the Court stated that '*the objective of PDOs and PGIs is to help producers of products linked to a geographical area by securing fair returns for the qualities of their products, by ensuring uniform protection of the names as an intellectual property right in the territory of the EU and by providing clear information on the value-adding attributes of the product to consumers*'. The use of the PDO 'Feta' to designate products produced in the territory of the EU '*which do not comply with the product specification for that PDO undermines those objectives, even if those products are intended for export to third countries. It therefore follows from the wording of Regulation No 1151/2012, as well as from its context and the objectives pursued by it, that such use constitutes conduct prohibited by that regulation*'.

The Court concluded that, by failing to prevent or stop such use in its territory, Denmark failed to fulfil its obligations under the EU GIs scheme.

---

[17] Commission Regulation (EC) No 1829/2002 of 14 October 2002 amending the Annex to Regulation (EC) No 1107/96 with regard to the name 'Feta', OJ 2002 L 277, p. 10.

[18] Op. cit 9, para. 82.

[19] C-159/20, Commission v Denmark (AOP Feta), ECLI:EU:C:2022:561, para. 63–66.

2.1.2 From *Rioja II* to *Schwarzwälder Schinken*: Shaping the Contours of the Internal Market

In the context of the absence of harmonisation, it was for the Court to conciliate the prohibition of any restriction to the free movement of goods and the protection of intellectual property rights at international level.[20] The confrontation between free movement of goods and GIs is similar to the one between this principle and other intellectual property rights (patents, trade marks, etc.).[21] Some opinions consider that the volunteer basis of the GIs scheme helps them to develop a peacefully existence if they do not interfere with the adequate interpretation of the free movement of goods and competition law.[22] Other views highlight the relevance of avoiding an unjustified fragmentation of the single market[23] or to reinforce the consumer protection and their participation in the scheme.[24]

Within the EU framework, the existence of different national legislations on GIs hindered the free movement of goods, one of the basic principles of the *acquis communautaire* and the creation of a common market, the EU created a sui generis system for GI protection, modelled on the existing national legislation.[25]

The compatibility with the internal market of a compulsory bottling in origin for PDO wine was the Gordian knot of *Rioja II*,[26] *Prosciutto di Parma*[27] and *Schwarzwälder Schinken.*[28]

In *Rioja II*, the Court declared that the GI's rules may restrict the free movement of goods when the measure is necessary for attainment of the objective pursued, the guarantee of the quality of the product, in that there are no less restrictive alternative measures capable of attaining it. In that case, the rule under review by the Court was Article 32 of the Rioja Wine standing orders, which states as follows:

> 1. Wine protected by the 'denominación de origen calificada' Rioja shall be bottled exclusively in the registered cellars authorised by the Governing Council, failing which the wine may not bear that designation.
>
> 2. Wines protected by the 'denominación de origen calificada' Rioja may be put into circulation and be despatched solely from registered cellars, in special bottles which do not detract from their quality or prestige and have been approved by the Governing Council. The bottles must be of glass and of a capacity authorised by the European Economic Community with the exception of one-litre bottle.

[20] Jarvis (1998), p. 295.

[21] Botana Agra (2001), p. 170.

[22] O'Rourke (1998), p. 72.

[23] Fuentes Nuñez (1995), p. 29.

[24] González Vaqué (1995), p. 60.

[25] Gragnani (2012), p. 272.

[26] C-388/95, Belgium v. Spain, ECLI:EU:C:2000:244, para. 76.

[27] Op. Cit 9.

[28] C-367/17, Schwarzwälder Schinken, ECLI:EU:C:2018:1025, para 35.

In a previous case *Delhaize*, identified as *Rioja I*,[29] the Court, in response to a request from the Tribunal de Commerce (Commercial Court, Brussels), ruled on the compatibility with Article 34 TFUE of national legislation (such as Decree No 157/88 and the Rioja wine standing orders adopted under it). It held that national provisions applicable to PDO wine which limited the quantity of wine that might be exported in bulk but otherwise permitted sales of wine in bulk within the region of production constituted measures having equivalent effect to a quantitative restriction on exports which were prohibited by Article 34 TFUE. In *Rioja I,* the Court then stated (paragraphs 22 and 23 of the judgment) that the justification claimed by the Spanish Government that the rules at issue formed part of a policy to improve the quality of wine could not be upheld since it had not been established that the confinement of bottling to a specified area could affect the quality of the wine. The reasoning behind of *Rioja I* was a clear statement in favour of the free movement of goods. When a wine entrepreneur invests in technological tools and human capacities for shipping the wine in raw to be bottled in the destination point and without altering their quality, it is difficult to understand why to impede him from doing it.[30]

As Spanish rules at issue in *Rioja I* were still in force, despite the interpretation of Article 34 TFUE given by the Court in that judgment, and the Commission did not issue a reasoned opinion, the Kingdom of Belgium commenced the Treaty-infringement proceedings which opened a second case, identified as *Rioja II.* In this second round of analysis of the Rioja Rules, the Court considered controls undertaken outside the region of production provide fewer guarantees as to the quality of the wine than those carried out in the region in compliance with the monitoring procedure and that: '.. *in the case of bottling, the best conditions are more certain to be assured if bottling is done by undertakings established in the region of those entitled to use the designation and operating under their direct control, since they have specialised experience and, what is more, thorough knowledge of the specific characteristics of the wine in question which must not be impaired or lost at the time of bottling'* (para. 65) and that '... *whilst it is true that the phenomenon of oxidation reduction may also occur during bulk transport in the region of production, even though the distance covered is usually less, restoration of the initial characteristics of the product will in those circumstances be entrusted to undertakings which offer all the guarantees of know-how and, again, the most thorough knowledge of wine'* (para. 66).

Therefore, the Court recognized that the measure was necessary for attainment of the objective pursued, in that there are no less restrictive alternative measures capable of attaining it as follows '...*the risk to which the quality of the product finally offered to consumers is exposed is greater where it has been transported and bottled outside the region of production than when those operations have taken place within the region'* (para 74) and '... *Accordingly, it must be accepted that the requirement at issue, whose aim is to preserve the considerable reputation of Rioja wine by strengthening control over its particular characteristics and its quality, is*

[29] C-47/90, Delhaize Frères v Promalvin and Others, ECLI:EU:C:1992:250, para. 22 and 23.

[30] Coello and González Botija (2005), p. 39.

justified as a measure protecting the 'denominación de origen calificada' which may be used by all the producers concerned and is of decisive importance to them'. (para. 75).

As in *Rioja II*, the proportionality principle in the context of the GI's is endorsed by the Court in the *Prosciutto di Parma* and *Schwarzwälder Schinken* judgments.

As explained in the precedent subsection, in *Prosciutto di Parma,* a condition such as at issue, in particular to the slicing and packaging of ham in the area of production, must be regarded as compatible with EU law, despite its restrictive effects on trade, if it was shown that it was necessary and proportionate and capable of safeguarding the quality of the product at issue, guaranteeing its origin or ensuring the control of the GIs specification.[31]

> 78. Consequently, the condition of slicing and packaging in the region of production, whose aim is to preserve the reputation of Parma ham by strengthening control over its particular characteristics and its quality, may be regarded as justified as a measure protecting the PDO which may be used by all the operators concerned and is of decisive importance to them (see, to that effect, Belgium v Spain, paragraph 75).
>
> 79. The resulting restriction may be regarded as necessary for attaining the objective pursued, in that there are no alternative less restrictive measures capable of attaining it.

The justification for making compulsory the 'sliced in origin' ham was considered proportional as no other measure could be considered comparable in terms of protection of the quality and the reputation of Parma ham because it was foreseen to strengthen the controls over characteristics and quality of the protected product.

> 80. The PDO Prosciutto di Parma would not receive comparable protection from an obligation imposed on operators established outside the region of production to inform consumers, by means of appropriate labelling, that the slicing and packaging has taken place outside that region. Any deterioration in the quality or authenticity of ham sliced and packaged outside the region of production, resulting from materialisation of the risks associated with slicing and packaging, might harm the reputation of all ham marketed under the PDO Prosciutto di Parma, including that sliced and packaged in the region of production under the control of the group of producers entitled to use the PDO (see, to that effect, Belgium v Spain, paragraphs 76 and 77).
>
> 81. Accordingly, the fact that the use of the PDO Prosciutto di Parma for ham marketed in slices is conditional on the slicing and packaging operations being carried out in the region of production may be regarded as justified, and hence compatible with Article 29 EC.

In *Prosciutto di Parma*, the Court concluded that the EU GIs system must be interpreted as not precluding the use of a PDO from being subject to the condition that operations such as the slicing and packaging of the product take place in the region of production, where such a condition is laid down in the specification.

In *Schwarzwälder Schinken*, it is said that it was for the national court to assess whether if the GIs specification constitutes a necessary and proportionate means to

[31] Op.cit 9, para. 66.

safeguard the quality of the product, to guarantee its origin or to ensure the verification of the specification of the GI. Moreover, the GI's rules may restrict the free movement of goods when the measure is necessary for attainment of the objective pursued, the guarantee of the quality of the product, in that there are no less restrictive alternative measures capable of attaining it. Proportionality principle in the context of the GI's is endorsed by the Court in *Prosciutto di Parma* and *Schwarzwälder Schinken*.[32]

2.2 An Ultimate Goal of the Common Agriculture Policy

2.2.1 Linked to the CAP Since 1992

In 1992, the legislator identified the GIs as part of the adjustment of the Common Agriculture Policy (CAP) and justified at that moment because of the need of encouraging the diversification of agricultural production as to achieve a better balance between supply and demand on the markets.[33]

In 2006, the reform focus was on the benefit to the rural economy, particularly in less-favoured or remote areas, as GIs were identified as a factor of improvement of the incomes of farmers and an instrument to retain the rural population in these areas and the possibility for the consumer to be able to make the best choices, by given clear and succinct information regarding the product origin.[34]

In 2012, the legislator inserted the GIs directly in the framework of the CAP and the Europe 2020 policy priorities; Regulation (EU) 1151/2012 was adopted.[35] Since that moment, the EU legislator undoubtedly stated the goals of the system into legal provisions.[36]

---

[32] Guillem Carrau (2022), p. 556.

[33] See second whereas of Council Regulation (EEC) No 2081/92 of 14 July 1992 on the protection of geographical indications and designations of origin for agricultural products and foodstuffs, OJ L 208, 24.7.1992.

[34] See whereas 2 and 4 of Council Regulation (EC) No 510/2006 of 20 March 2006 on the protection of geographical indications and designations of origin for agricultural products and foodstuffs, OJ L 93, 31.3.2006.

[35] Regulation (EU) No 1151/2012 of the European Parliament and of the Council of 21 November 2012 on quality schemes for agricultural products and foodstuffs, OJ L 343, 14.12.2012.

[36] Article 1 Regulation (EU) 1151/2012, where it was said that GIs system aimed '*…to help producers of agricultural products and foodstuffs to communicate the product characteristics and farming attributes of those products and foodstuffs to buyers and consumers, thereby ensuring: (a) fair competition for farmers and producers of agricultural products and foodstuffs having value-adding characteristics and attributes; (b) the availability to consumers of reliable information pertaining to such products; (c) respect for intellectual property rights; and (d) the integrity of the internal market. With the same approach, Article 4 Regulation (EU) 1151/2012 said that 'A scheme for PDO and PGI is established in order to help producers of products linked to a geographical area by: (a) securing fair returns for the qualities of their products; (b) ensuring uniform protection of the names as an intellectual property right in the territory of the Union; (c) providing clear information on the value-adding attributes of the product to consumers*'.

The goal of the legislation on PDOs and PGIs is therefore to protect traditional products 'with specific characteristics linked to geographical origin'. As stated by the Advocate General in *Scotch Whisky Association*,[37] the protection granted to those terms merely serves to fulfil that objective, and the scope of such protection must therefore be interpreted in the light of that objective.

> 37. With regard to the objectives pursued, in particular, by Article 16 of Regulation No 110/2008, it is clear from, inter alia, its title that the objective of that article is to ensure the 'protection of geographical indications' through a registration of those indications which aims, on the one hand, to allow the identification of spirit drinks as originating in a specific territory in the situations set out in recital 14 and, on the other, to contribute to the achievement of the more general objectives set out in recital 2.

In 2024, the new EU GIs scheme reinforces the link to CAP as it can be gathered from some of the recitals of Regulation (EU) 2024/1143 and, specially, from the statements contained in recital 19 where it is said that quality schemes are able to contribute to and complement rural development policy as well as market and income support policies of the CAP. In particular, the legislator highlights that GIs can contribute to developments in the farming sector and, especially, disadvantaged areas and connects the GIs with the Commission communication of 30 June 2021 entitled 'A long-term vision for the EU's Rural Areas — Towards stronger, connected, resilient and prosperous rural areas by 2040'. This document recognised the key role of GIs among the flagship initiatives that promote rural areas, in view of their contribution to the prosperity, economic diversification and development of rural areas and the strong association between a product and its territorial origin. Likewise, the legislator points out that a Union framework that protects GIs by providing for their inclusion in a register at Union level facilitates the development of the agricultural sector, since the resulting, more uniform approach ensures fair competition between the producers of products bearing such indications and enhances the credibility of products from the perspective of consumers.

Definitively, GIs are a privileged tool to work for local and rural development.[38] GIs perfectly connect producers to consumers whose aims are an agricultural sector less subsidized and more diverse in the EU.[39] They also protect consumers' interests by preventing practices liable to mislead consumers.[40]

2.2.2 *Mozzarella di Bufala* Case as Snapshot of Legislative Objectives

Mozzarella di Bufala Campana, clarifies that the main purpose of the EU GIs scheme is related to the support of producers. Verbatim, it is said that '… *provides*

[37] C-44/17, Scotch Whisky Association, AG Conclusions, ECLI:EU:C:2018:111, para. 37.

[38] Berengue (2004).

[39] Addor and Grazioli (2002), p. 874; Babcock and Clemens (2004), p. 17; Lall (2003), p. 6.

[40] Montero (2021), p. 427.

that a scheme for PDOs and PGIs is to be established in order to help producers of products linked to a geographical area by providing clear information on the value-adding attributes of the products to customers'.[41]

In this context, the General Court also defined the aim of the scheme as to guarantee the reputation of the protected products amongst the consumers:

> 36. …The applicable rules protect those entitled to use them against improper use of those designations by third parties seeking to profit from the reputation which they have acquired. They are intended to guarantee that the product bearing them comes from a specified geographical area and displays certain particular characteristics. They may enjoy a high reputation amongst consumers and constitute for producers who fulfil the conditions for using them an essential means of attracting custom. The reputation of designations of origin depends on their image in the minds of consumers. That image in turn depends essentially on characteristics and more generally on the quality of the product. It is on the latter, ultimately, that the product's reputation is based. For consumers, the link between the reputation of the producers and the quality of the products also depends on his being assured that products sold under the designation of origin are authentic.

To sum up, the EU GIs system relies on the following legislative objectives:

- to help producers of products linked to a geographical area by providing clear information on the value-adding attributes of the products to customers.
- to secure higher incomes in return for a genuine effort to improve quality.
- to protect those entitled to use GIs against improper use of those designations by third parties seeking to profit from the reputation which they have acquired. They are intended to guarantee that the product bearing them comes from a specified geographical area and displays certain particular characteristics.

2.2.3 A Set of CAP Objectives that Inspired Article 4 Regulation (EU) 2024/1143

Recital 19 of Regulation (EU) 2024/1143 sums up the goals detailed in Article 4 Regulation (EU) 2024/1143 in a concise paragraph where the legislator synthetizes the real aim of the EU quality food system. The text is self-explanatory of the legislator view. In the context of the CAP, GIs schemes are '*crucial to tackle a scenery of collective challenges: higher sales premium for protected products that will help to ensure fair distribution of the value added amongst the actors in the supply chain; access to promotion funds, investment and rural development support and creation of synergies in GIs territory that will complement the rural development policy; and of course, connect with consumer and help them to take better decisions when purchasing food*'.

As a matter of fact, Article 4 Regulation (EU) 2024/1143 declares that it provides for a unitary and exhaustive system of GIs by the definition of the following goals:

---

[41] C-569/18, Caseificio Cirigliana and Others, ECLI:EU:C:2019:873, para. 32.

(a) ensuring that producers acting collectively have the necessary powers and responsibilities to manage the geographical indication concerned, including in order to respond to societal demands, such as for animal health and welfare, for products resulting from sustainable production in its three dimensions of economic, environmental and social value, and to operate and be competitive in the market;
(b) contributing to fair competition and generating added value with the aim of sharing that added value across the marketing chain, in order to ensure a fair return for producers and a capacity to invest in the quality, reputation and sustainability of their products, as well as contributing to the achievement of rural development policy objectives by providing support to agricultural and processing activities, preserving know-how and promoting specific quality products due to the geographical area where they are produced;
(c) ensuring that consumers receive reliable information and necessary guarantee of the origin, authenticity, quality, reputation and other characteristics linked to the geographical origin or the geographical environment of such products and can readily identify them in the marketplace including in electronic commerce;
(d) ensuring the efficient and user-friendly registration of geographical indications taking into account the appropriate protection of intellectual property rights;
(e) ensuring effective controls, enforcement and placing on the market throughout the Union, including in electronic commerce, thereby ensuring the integrity of the internal market; and
(f) contributing to the effective protection of intellectual property rights related to such products in third country markets.

Currently this legal provision is defining the common scope of protection for wine GIs [according to the new content of Article 103 (Regulation (EC) 1308/2013[42])]; spirit drinks GIs (as Article 21 Regulation (EU) 2019/787 is deleted following Article 85 of Regulation (EU) 2024/1143[43]); and for aromatised wine product with geographical indication (as Article 20 Regulation (EU) No 251/2014 was deleted by Regulation (EU) 2021/2117).[44]

2.3 An Indivisible Part of the Common Commercial Policy: GIs as a *hot potato* in Trade Agreements

At international level, several treaties administered by the World International Property Organisation (WIPO) provide for the protection of GIs, most notably the Paris Convention for the Protection of Industrial Property of 1883,[45] the Madrid

[42] Article 103 says that 'Protected designations of origin and protected geographical indications referred to in this Regulation shall be protected in accordance with Articles 26 to 31, 35 and 36 of Regulation (EU) 2024/1143'. OJ L 347, 20.12.2013.

[43] OJ L 130, 17.5.2019.

[44] OJ L 435, 6.12.2021, pp. 262–314.

[45] GI protection was part of the Paris Convention for the Protection of Industrial Property (1883), but under a different label ('false indication'). Article 10 of the Paris Convention refers generally to "indications of source" and "appellations of origin" as objects of industrial property, but has not a definition of these concepts. The direct or indirect use of a false indication of source of a product or of a producer, processor or trader is forbidden under this Convention. However, the relevancy of

Agreement for the Repression of False or Deceptive Indications,[46] and the Lisbon Agreement for the Protection of Appellations of Origin and their International Registration.[47] In addition, Articles 22–24 of the Agreement on Trade-Related Aspects of Intellectual Property Rights (TRIPS) deal with the international protection of GIs within the framework of the World Trade Organization (WTO).[48] The above-mentioned international agreements allow for the granting of GI protection to all product types, both agricultural and non-agricultural.

At this level—and we are not considering bilateral agreements—the first international agreement that gave a definition of *appellations of origin* was the Lisbon Agreement for the Protection of Appellations of Origin and their International Registration, in 1958. Within WIPO, Geneva Act was essentially intended to facilitate and govern trade between the Union and the third States party to the Lisbon Agreement.

Falling at the confluence of agriculture, trade and intellectual property, GIs have become 'a red-hot issue' in international law.[49]

2.3.1 Commission v Council (Revised Lisbon Agreement): Who Negotiates on Behalf of the EU

In November 2019, the EU acceded to the Geneva Act of the Lisbon Agreement and permitted its entry into force in 2020. Regulation (EU) 2023/2411 opened the

---

this Convention is that the TRIPS Agreement imposes Members of the WTO to comply with Articles 1 through 12, and Article 19, of the Paris Convention (after its revision in 1967).

[46] The 1891 Madrid Agreement for the Repression of False or Deceptive Indications also addressed GIs.

[47] By first time, an International Agreement contained a definition of Appellation of Origin and protected the sign from any imitation or misused and any non-authorised, direct, or indirect, uses. This Agreement inspired several bilateral Agreements and the EU GIs scheme (Regulation (EEC) no 2081/1992).

[48] During the Uruguay Round, with the 1995 multilateral TRIPs Agreement, GIs were introduced for the first time into international trade treaties by setting the minimum standards that every WTO Member State must respect. Article 22 TRIPS sets forth a definition of a GI and contains a general obligation for WTO Members to provide protection against misleading use of a GI and against use that constitutes an act of unfair competition. It also requires Parties to refuse or invalidate the registration of trade marks that contain or consist of a GI with respect to goods not originating in the territory indicated, if use of the indication on the trade mark for such goods might mislead the public as to the true place of origin. Article 23 TRIPS accords a higher level of protection to wine and spirits GIs. Article 23(1) sets a standard of absolute protection which requires Parties to prevent a registration of a wines/spirit's GI on wines or spirits not originating in the identified geographical location, even if the true origin of the goods is indicated or the geographical indication is used in translation or accompanied by expressions such as "kind", "type", "style", "imitation" or the like. Wines and spirits GIs have to be protected even if misuse of a GI name would not mislead the consumers.

[49] Raustiala and Munzer (2007), p. 342.

possibility to get protection under the Geneva Act for EU producers. Respectively, the EU will be accepting requests for such protection from members of the Geneva Act.[50]

The EU has exclusive competence for the areas covered by the Geneva Act. This was confirmed in *Council vs Commission (revised Lisbon Agreement*[51]*)* which clarified that the draft revised Agreement, which has been subsequently adopted as the Geneva Act, concerned with commercial aspects of intellectual property and hence fall within the Common Commercial Policy and not in the field of the internal market as sustained by the Council and some Member States[52]:

> 74 It thus follows from the examination of the draft revised agreement, first, that it is essentially intended to facilitate and govern trade between the European Union and third States and, secondly, that it is such as to have direct and immediate effects on such trade, so that its negotiation falls within the exclusive competence which Article 3(1) TFEU confers on the European Union in the field of the common commercial policy envisaged in Article 207(1) TFEU.
>
> 75 Therefore, the Council was wrong in taking the view that the contested decision fell within the approximation of laws in the field of the internal market and, accordingly, within a competence shared between the European Union and its Member States, and in basing that decision on Article 114 TFEU and Article 218(3) and (4) TFEU.

All past agreements relating to GIs had been exclusively based on Article 207 TFEU and the case-law confirms this interpretation.[53] The Court confirmed, in the referred case, that the aim and objective of the Lisbon Agreement was therefore to protect GIs in countries other than the producing country. This only made sense in view and with the objective of promoting, facilitating and governing trade between these countries in these products.

> 76 Contrary to the Council's contentions, that error cannot be regarded as a mere formal defect. In particular, it led the Council to disregard the procedural provisions specifically laid down in Article 207(3) TFEU for the negotiation of international agreements falling

[50] Article 64 Regulation (EU) 2023/2411 contains the necessary amendments to Article 1 and 2 Regulation (EU) 2019/1753 to empower the Commission or the EUIPO to file applications for the international registration of GIs protected and registered under Union law and pertaining to products originating in the Union pursuant to Article 5(1) and (2) of the Geneva Act with the International Bureau of the World Intellectual Property Organization ("the International Bureau") and proceed on the basis of the decision on granting protection. Besides, it is foreseen that EUIPO will publish any international registration notified by the International Bureau pursuant to Article 6(4) of the Geneva Act and assess it. Likewise, EUIPO will grant the possibility of lodging opposition and admit or reject it and decide to grant the protection of the GI. If needed, EUIPO will notify the International Bureau of the refusal of the effects of the international registration concerned in the territory of the Union.

[51] C-389/15, Council vs. Commission, ECLI:EU:C:2017:798, para. 74.

[52] See, as regards rules and procedures concerning actions of the Union as regards the accession, Regulation (EU) 2019/1753, OJ L 271, 24.10.2019.

[53] C-347/03, Regione autonoma Friuli-Venezia Giulia and Agenzia regionale per lo sviluppo rurale (ERSA), EU:C:2005:285.

> within the field of the common commercial policy, above all those relating to conduct of the negotiations by the Commission, as the Advocate General has observed in points 86 and 89 of his Opinion.

The Lisbon Agreement had thus immediate effects on trade, said the Court. It therefore falls within the common commercial policy. The Lisbon Agreement relates specifically to trade as requested by the Court most recently in the *Daiichi Sankyo* case.[54]

> 46. For its part, the Commission's approach is that the wording of Article 207(1) TFEU constitutes a reference, implicit but no less clear, to the subject-matter governed by the TRIPs Agreement: such a striking parallelism between the wording of Article 207 TFEU and the title of the agreement practically leads to that conclusion. Essentially, its argument is simply that it is impossible to imagine that the European Union legislature had any intention other than that of raising to the status of exclusive competence of the European Union an area of substantive law, the 'commercial aspects of industrial property', which was provided for in a very different way by Article 133 EC[55]....

2.3.2 An Extension of the EU GIs Scope of Protection Under the Geneva Act

The entry into force of the Geneva Act of the Lisbon Agreement will manage to bridge the divide between the countries that promote the *sui generis* model of protection of GIs[56] and provides a stronger protection to GIs that the EU legal framework.

Upon the accession of the Union to the Geneva Act and thereafter on a regular basis, the Commission should file with the International Bureau of the World

---

[54] C-414/11—Daiichi Sankyo and Sanofi-Aventis Deutschland, ECLI:EU:C:2013:520, para 46.

[55] The Judgment in para. 46 has the following footnote: '*As compared to the former Article 133 EC (post Nice), Article 207 TFEU differs by specifically including 'the commercial aspects of intellectual property' among the subject-matters in respect of which '[t]he common commercial policy shall be based on uniform principles', whereas Article 133(5) EC, simply provided that Paragraphs 1 to 4 were also to apply 'to the negotiation and conclusion of agreements in the fields of trade in services and the commercial aspects of intellectual property, in so far as those agreements [we]re not covered by the said paragraphs', the first of which required that the common commercial policy was to be based on uniform principles. Article 133 EC therefore already covered, in that field and as a matter of principle, the external dimension of the commercial aspects of intellectual property. Article 207 TFEU simply includes them in a direct and comprehensive manner, beyond the mere external dimension. In line with this, Advocate General Kokott considered, in point 63 of her Opinion in Case C 13/07 Commission v Council, that pursuant to Article 133(5) EC the Community had not acquired 'exclusive competence ... in the field of trade in services and the commercial aspects of intellectual property', but rather that 'that step [was] completed only in the Treaty of Lisbon: Article 207(1) TFEU henceforward expressly places the "new" fields of commercial policy on the same footing as the conventional fields, and the common commercial policy as a whole is expressly assigned to the exclusive competence of the Union (Article 3(1)(e) TFEU)'. This does not mean, however, that Article 207 TFEU has conferred on the European Union exclusive competence for intellectual property law*'.

[56] Lightbourne (2021), p. 932.

Intellectual Property Organization (WIPO) applications for the international registration of GIs originating and protected in the territory of the Union in the register of the International Bureau. This action will be implemented according to the rules of Regulation (EU) 2019/1753, of 23 October 2019, on the action of the Union following its accession to the Geneva Act of the Lisbon Agreement on Appellations of Origin and Geographical Indication, which lays down rules allowing the Union to exercise the rights and to fulfil the obligations laid down in the Geneva Act.[57]

The Geneva Act protects appellations of origin, including designations of origin within the meaning of EU Regulations. Regarding the GIs' protection, Article 11(1) (a) of Geneva Act stipulates that the Contracting Parties of the Lisbon Agreement, shall provide the legal means to prevent:

> (a) use of the appellation of origin or the geographical indication
>
> (i) in respect of goods of the same kind as those to which the appellation of origin or the geographical indication applies, not originating in the geographical area of origin or not complying with any other applicable requirements for using the appellation of origin or the geographical indication;
>
> (ii) in respect of goods that are not of the same kind as those to which the appellation of origin or geographical indication applies or services, if such use would indicate or suggest a connection between those goods or services and the beneficiaries of the appellation of origin or the geographical indication, and would be likely to damage their interests, or, where applicable, because of the reputation of the appellation of origin or geographical indication in the Contracting Party concerned, such use would be likely to impair or dilute in an unfair manner, or take unfair advantage of, that reputation;
>
> (b) any other practice liable to mislead consumers as to the true origin, provenance or nature of the goods.
>
> (2) Paragraph (1)(a) shall also apply to use of the appellation of origin or geographical indication amounting to its imitation, even if the true origin of the goods is indicated, or if the appellation of origin or the geographical indication is used in translated form or is accompanied by terms such as "style", "kind", "type", "make", "imitation", "method", "as produced in", "like", "similar" or the like.

As reproduced, the provision refers to "*any direct or indirect use … when products are comparable, or where using the name exploits the reputation of the protected name*", indicating that not only use for comparable products is prohibited. In addition, the other subparagraphs, including (b) ("evocation"), which is not subject to a comparability criterion.

Under the Geneva Act, there is an absolute protection of the GI for goods of the same kind as those to which the GI refers. In respect of goods that are not of the same kind as those to which the appellation of origin or geographical indication applies or services, if such use would indicate or suggest a connection between those goods or services and the beneficiaries of the appellation of origin or the geographical indication, and would be likely to damage their interests, or, where applicable, because of the reputation of the appellation of origin or geographical indication in the Contracting Party concerned, such use would be likely to impair or dilute in an unfair manner, or take unfair advantage of, that reputation.

[57] OJ, L 271, 24.10.2019.

Therefore, it could have been interpreted that the Geneva Act provides stronger protection to GIs than the EU legal framework, which could be understood as limited to the comparable ones in the past and could be quoted in Court in the interest of any concerned GI. Furthermore, it can be considered the Geneva Act as being directly applicable when, regard being had to its wording and the purpose and nature of the agreement itself, their provisions contain a clear and precise obligation which is not subject, in its implementation or effects, to the adoption of any subsequent measure.[58] Currently, this interpretation has been clarified after the last review of the EU scheme: Article 26 Regulation (EU) 2024/1143 also includes other goods and services: "*... or where use of that geographical indication for any product or any service exploits, weakens, dilutes, or is detrimental to the reputation of, the protected name, including when those products are used as an ingredient*".

Likewise, the Geneva Act means expanding the protection of the GIs against any unauthorized use in respect of goods that are not of the same kind as those to which the GI applies, or services, if such use would indicate or suggest a connection between those goods or services and the beneficiaries of GI.

As it is known, international agreements with international organisations are an integral part of EU law. These agreements are separate from primary law and secondary legislation and form a sui generis category. According to some case law, they can have direct effect, and their legal force is superior to secondary legislation, which must therefore comply with them. A provision in an agreement concluded by the Union with non-member countries must be regarded as being directly applicable when, regard being had to its wording and the purpose and nature of the agreement itself, the provision contains a clear and precise obligation which is not subject, in its implementation or effects, to the adoption of any subsequent measure.[59]

Consequently, the Geneva Act is to be considered, as far as the Union is concerned, an act of one of the EU institutions within the meaning of point (b) of the first subparagraph of Article 267 TFEU, and, as from its entry into force, the provisions of such an agreement form an integral part of the Union legal system and, within the framework of that system, the Court has jurisdiction to give preliminary rulings concerning the interpretation of such an agreement.[60]

2.4 Conclusions

In the EU, GIs are underpinned by three fundamental pillars: the common agriculture policy, the internal market, and the common commercial policy, as established by case law. Based on the detailed analysis of key cases such as *Feta, Prosciutto, Mozzarella, Rioja II, Schwarzwälder Schinken*, and *Commission v Council (Revised Lisbon Agreement)*, this chapter makes a set of technical findings that are useful to understand how the main objective of the so-called EU GIs scheme is to promote

[58] Guillem Carrau (2022), p. 564.

[59] C-12/86, Demirel v Stadt Schwäbisch Gmünd, EU:C:1987:400, para 14.

[60] C-181/73, Haegemann v Belgian State, EU:C:1974:41, para. 3–6.

quality, genuine character, variety, and value in the food chain, together with the preservation of traditional know-how in the sector and to stimulate diversification and employment in rural areas. By conferring unitary protection on GIs, the EU GIs scheme provides a robust framework for safeguarding the authenticity of the European agriculture, thereby upholding the interests of both producers and consumers alike in the Internal market, including the new digital environments. Finally, our analysis helps to reach logical conclusions about the fact that GIs have begun to assume a pro-active and increasingly collective bargain role in international trade negotiations and, consequently, will be benefiting directly from their results, as the Geneva Act example shows, by granting a stronger scope of protection.

References

Addor F, Grazioli A (2002) Geographical indications beyond wines and spirits: a roadmap for a better protection for geographical indications in de WTO TRIPS Agreement. J World Intellect Prop 5(6)

Audier J (2003) Quelle stratégie juridique pour la commercialisation du produit agricole ? (Which legal strategy for marketing the agricultural product?) Marques et indications géographiques de la filière vitivinicole. Revue de Droit Rural, n° 311 mars

Babcock B, Clemens R (2004) Geographical indications and property rights: protecting value-added agricultural products. Matric briefing paper 04-MBP 7, Mayo 2004

Beier FK, Knaak R (1994) The protection of the direct and indirect geographical indication of source in Germany and in the European Community. IIC international review of industrial property and copyright law

Berengue A (2004) Geographical origins in the world. Paper presented at the workshop "Promoting agricultural competitiveness through local know-how". Proceedings of the Montpellier Workshop. World Bank Group/MAAPAR/CIRAD, Washington DC/Paris/Montpellier

Botana Agra M (2001) Las Denominaciones de Origen, Tratado de Derecho Mercantil, tomo XX, vol.2°, Marcial Pons, Madrid.

Brouwer O (1991) Community protection of geographical indications and specific character as a means of enhancing foodstuff quality. Common Mark Law Rev 28:615–646

Calboli I (2017) Geographical indications between trade, development, culture, and marketing: framing a fair(er) system of protection in the global economy? In: Calboli I, Ng-Loy WL (eds) Geographical indications at the crossroads of trade, development, and culture. Cambridge University Press, Cambridge, pp 3–35

Coello C, González Botija F (2005) El embotellado en destino del vino de Jerez (Sherry wines and their bottling in destination). Revista de Derecho Agrario y Alimentario, 2° época, n° 45, año XXI

Favret JM (2000) Droit Communautaire du marché interieur (EU internal market law), Gaulino edn. Paris

Fuentes Nuñez LA (1995) La protección de las denominaciones de origen en derecho comunitario (The GIs protection in EU law). Gaceta Jurídica de la CEE, B-101

Gangjee D (2012) Relocating the law of geographical indications. Cambridge University Press, Cambridge

González Sacedón J, Vassilakis G (1998) La denominación de origen en derecho comunitario (The PDO in EU law). Paper presented in Jornadas sobre denominaciones de origen, Diputación General de Aragón

González Vaqué L (1995) "Tribunal de Justicia de las Comunidades Europeas: la nueva jurisprudencia en materia de libre circulación de mercancías", Estudios sobre Consumo, n° 34.

Gragnani M (2012) The law of geographical indications in the EU. J Intellect Prop Law Pract 7(4):271–282
Guillem Carrau J (2022) How much protection for geographical indications (GIs)? A jigsaw of case law is falling into place in the EU. EIPR 44(9):556–566
Guillem Carrau J (2023) Towards the extension of geographical indication protection of the European Union beyond the agricultural products: an EU craft GI's system. EIPR 45(12):725–736
Jarvis MA (1998) The application of EC law by national courts: the free movement of goods (Oxford, online edn, Oxford Academic, 31 Oct 2023)
Kur A (2023) No strings attached to GIs? About a blind spot in the (academic) discourse on limitations and fundamental rights. IIC 54:87–94
Kur A, Cocks S (2007) Nothing but a GI thing: geographical indications under EU law. Fordham Intellect Prop Media Ent Law J 17:999–1016
Lall CHM (2003) Geographical indications in developing countries. Special report on GI, September, INTA
Lightbourne M (2021) What's in a name? The journey of geographical indications from Paris 1883 to Geneva 2015. GRUR Int 70(10):932–942
Mattera A (1991) El Mercado Único. Sus reglas. Su funcionamiento (Internal market: rules and development), Madrid
Montero P (2021) Towards a core unitary legal regime for geographical indications in the European Union digital market. J Intellect Prop Law Pract 16(4-5):427–434
Oliver P (1996) Free movement of goods in the European Community under articles 30 to 36 of the Rome Treaty. Sweet & Maxwell, London
O'Rourke R (1998) European Food Law, Palladian Law Publishers, Brussels.
Raustiala R, Munzer S (2007) The global struggle over geographic indications. EJIL 18(2):337–365
Swann D (1992) The economics of the common market, 7th edn. Penguin

The Key Elements of the EU Geographical Indications System 3

Abstract

A jigsaw of case law helps to understand what is under protection (*Warstein, Pistre and Haus Cramer*) and identifies a set of components of the EU GIs scheme: their place into the intellectual property rights and the existence of an EU title which is obtained via Register (*Rioja II, Turrón de Jijona and Salame felino*) and whose exclusivity prevents Member States from create parallel schemes, as said in *Budějovický Budvar beer*; the fact that qualities and characteristics are due to the specific geographical area of origin, as described in *Sekt*; the GIs intrinsic reputation as discussed in the decision *Nero Champagne*; the value of the legislative position and its historical development in *Grana Biraghi*; the protection of the terroir and the savoir faire of the product, associated to its origin, that can benefit only the eligible producers, as declared in *Manchego*; and the fact that generic terms are not protected because the protection is reserved to the protected term as such, as clarified in *Aceto Balsamico di Modena*. The last component under review is the concept of sustainability.

In this chapter, with the help of cases like *Rioja II* and *Turrón de Jijona* we will describe how GIs are a family member of the intellectual property rights or, at least, a new kid on the block. Other judgments as *Budějovický Budvar beer* will show a clear commitment on the exclusivity of the EU GIs scheme that prevents Member States from create parallel ones. *Sekt* case will demonstrate that the EU GIs system relies on the fact that qualities and characteristics are due to the specific geographical area of origin and, following *Manchego cheese* case, that the protection of the *terroir* and the *savoir faire* of the product, associated to its origin, can benefit only the eligible producers. For PGIs the protection can be obtained on the singular justification of the reputation of the product, as referred in *Nero Champagne* Boards of Appeal's Decision. In *Grana Biraghi*, it is shown how GIs protection depends on

© The Author(s), under exclusive license to Springer Nature Switzerland AG 2026

J. Guillem Carrau, *Landmark EU Geographical Indications Case Law*, Law for Professionals, https://doi.org/10.1007/978-3-032-26101-4_3

the legislative protection and its historical development and not on definitions of the dictionaries or internet extracts. Another key element is that generic terms are not protected. The protection is reserved to the protected term as such, as clarified in *Aceto Balsamico de Modena*. Ultimately, Regulation (EU) 2024/1143 contains a new element related to sustainability of the system that has been inserted directly in the scheme following the political discussion on the Green Deal.[1]

At this point, as a preliminary remark on terminology, even though we use the term GIs for both PDOs and PGIs, it is relevant to describe how the EU legislator distinguishes these two categories PDOs and PGIs in Article 46 Regulation (EU) 2024/1143. As it can be gathered from the legal provision, regarding the PDO, all the production steps will take place in the defined geographical area:

Article 46

1. A 'designation of origin' of an agricultural product is a name which identifies a product:

(a) originating in a specific place, region or, in exceptional cases, a country;

(b) whose quality or characteristics are essentially or exclusively due to a particular geographical environment with its inherent natural and human factors; and

(c) the production steps of which all take place in the defined geographical area.

2. A 'geographical indication' of an agricultural product is a name which identifies a product:

(a) originating in a specific place, region or country;

(b) whose given quality, reputation or other characteristic is essentially attributable to its geographical origin; and

(c) at least one of the production steps of which takes place in the defined geographical area.

From this perspective, on the one hand, the legal categories created by the legislator have managed to offer a common and binding framework for promoting the quality of European products, but the complexity of this makes it difficult for the average consumer to understand the characteristics of the product being offered and find a homogeneous level of quality in it. On the other, the existence of two categories, PDOs and PGIs, and a single distinctive graphic sign seems to have facilitated the acceptance of these by the consumer and the automatic assignment of good will.[2]

Most of the GI critics are focused on the anticompetitive and protectionist character of this system.[3] Kur considers that no relationship between quality and *terroir*

---

[1] Recital 3 Regulation (EU) 2024/1143.

[2] Guillem Carrau (2021), p. 293.

[3] O'Rourke (1998), p. 77.

must be established for PGIs; it suffices that the geographic origin accounts for the reputation of the goods. Furthermore, of three steps—production, processing and preparation—only one must be performed in the designated area, while the locality requirement applies to all three steps in case of PDOs. Despite the lower threshold thus established, PGIs enjoy the same protection as PDOs, thereby devaluating the original link between requirements and scope of the exclusive right. Furthermore, less obvious than by the inclusion of PGIs but equally efficient, the notion of strictly terroir-based protection has been undercut by increasing emphasis being placed on human (instead of natural) factors accounting for the specific quality of products designated by a PDO. There is at least a silent understanding among those dealing with the subject that today, protection under the EU sui generis regime—including protection of PDOs—is more about tradition than terroir.[4]

In fact, this formal differentiation but substantial equalization between PDOs and PGIs has always been criticized as the real contradiction—if not the 'original sin' in the legislative compromise—of the EU system,[5] providing absolute protection against parasitism, beyond the standards of unfair competition for misleading practices, even to products which present a looser qualitative-territorial link in terms of reputational requirements, production steps and raw materials.[6]

3.1 The *Rioja II* and the *Turrón* de *Jijona*: A GI Constitutes an Intellectual and Commercial Property Right

Within the EU framework, the existence of different national legislations on GIs hindered the free movement of goods, one of the basic principles of the *acquis communautaire* and the creation of a common market, the EU created a sui generis system for GI protection, modelled on the existing national legislation.[7]

In this context, as stated by the European Court of Justice in *Turrón de Jijona*,[8] also known as *Exportur*, a GI constitutes an intellectual and commercial property right; verbatim '*...may be regarded as falling within the sphere of the protection of industrial and commercial property within the meaning of Article 36...*'

The reference for a preliminary ruling was concerned with the lawfulness under EU law of protection of geographical names by Member States. The respondents in the main proceedings, the French undertaking LOR SA and Confiseries du Tech, manufactured and distributed in France a range of confectionery products which include 'tourons', a type of nougat, under a Spanish name. LOR SA's products included the 'tourons Alicante' and 'tourons Jijona' and the products of Confiseries du Te ch included 'tourons type Alicante' and

[4] Kur (2023), p. 89.

[5] Calabrese (2024), p. 693.

[6] Gangjee (2012), p. 211.

[7] Gragnani (2012), p. 272.

[8] C-3/91, Exportur, ECLI:EU:C:1992:420, para. 37. See, as regards the litigation in France: Gaultier (2005), p. 501.

'tourons type Jijona'. The appellant, a Spanish association of firms exporting tourons from Jijona, brought an action against the two respondent firms for an injunction restraining them from manufacturing and marketing nougats (*tourons*) in France under the names 'Alicante' and 'Jijona'. It relied on the Franco-Spanish Convention of 1973.[9] Pursuant to Article 3 of this Convention, in conjunction with Annex B thereto, the names, inter alia, 'Turrón Alicante' and 'Turrón de Jijona' were (and are now under EU Law) reserved exclusively for Spanish products or Spanish goods and might be used only in accordance with the conditions laid down in the legislation of the Spanish State. The respondents on the other hand took the view that after Spain's accession the Convention infringed Community law and the appellant could therefore not rely on it.

In *Jijona* case, AG Opinion declared that, only if the indications or designations in question were generic names, the protection of the Convention was not acceptable under EU Law. In the same vein, the Court stated that the Treaty do not impede the application of a bilateral convention because the names at hand were not generics in Spain:

> Articles 30 and 36 of the Treaty do not preclude the application of rules laid down by a bilateral convention between Member States on the protection of indications of provenance and designations of origin, such as the Franco Spanish Convention of 27 June 1973, provided that the protected names have not, either at the time of the entry into force of that Convention or subsequently, become generic in the country of origin.

Another clear message about the nature of GIs as intellectual property rights can be found in *Rioja II*:

> ... Designations of origin fall within the scope of industrial and commercial property rights. The applicable rules protect those entitled to use them against improper use of those designations by third parties seeking to profit from the reputation which they have acquired. They are intended to guarantee that the product bearing them comes from a specified geographical area and displays certain particular characteristics.[10]

GIs' legal environment is rooted to a specific and different right in the field of the intellectual property or, an intellectual collective property right of special category.[11] The patrimonial element is differently reflected in GIs as regards other types of intellectual property rights because GIs cannot be sold, transferred, or licensed. They can be used only if the conditions of the single documents are respected.[12]

In this context, there are two basic problems, firstly geographical terms a priori do not work as trade marks because they are descriptive and not distinctive and,

---

[9] See Article 5(1) of the Convention that requires the use of names in contravention of the Convention to be suppressed; according to Article 5(2), the same requirement applies even where the name is translated or indicates the true provenance or is accompanied by terms such as 'style', 'kind', 'type', 'imitation' or 'similar'.

[10] C-388/95, Belgium v. Spain, ECLI:EU:C:2000:244, para. 54.

[11] Tinlot et al. (1987), p. 32.

[12] Pérez Ferreras (2004), p. 14.

secondly, goods and services identified with the geographical terms must come from the geographical origin or they will be considered misleading.[13]

This reflection is connected to *Turrón de Jijona* case where the Court stipulated that they may enjoy a high reputation amongst consumers and constitute for producers who fulfil the conditions for using them an essential means of attracting custom.[14]

Notably, the new unified regime for agri-food and wine products has incorporated the previous one for spirits which, as known, provides for just one single form of GI, substantially coinciding with PGI, and likewise, the new parallel scheme for craft and industrial products provides for only a single form of GI, again modelled on PGI.[15]

In fact, GIs have a difficult insertion within the classic intangible rights and, nevertheless, it aggregates elements from those intangible rights to create a singular concept within the orbit of the Commercial Law, and particularly the Intellectual Property. This highlights the need of improvement as regards the so-called interconnecting or conflict rules.[16]

As quoted in Chap. 1, one needs to consider that GIs are a "special animal" in terms of their substance and history; they cannot simply be put on equal footing with any other IP right. At their root lies the concept of *terroir* giving certain products their unique characteristics and quality due to the natural surroundings. This notion gave GIs their special place in the universe of IP law and justified their broad protection.[17]

3.2 A Light of Clarity on the Terminology: What Is Under Protection? (*Warstein, Pistre and Haus Cramer*)

The AG Opinion in *Warstein* helps to identify which indications of origin are protected under the EU GIs legal framework by discarding geographical indications of source. As the AG Opinion indicates '*...the terminology used in this area itself risks being a fruitful source of confusion. I will use the term 'geographical indication of source' in the sense simply of a place name designating a product where, first, there is no link between the characteristics of the product and its geographical provenance and, secondly, use of the name may or may not lead consumers to think that the product originates in that place; it is clear from the order for reference that this is moreover the sense in which the referring court uses the term 'simple geographical indication of source'. I will in contrast use the terms 'geographical indication' and 'designation of origin' only in the sense in which they are defined in the EU GIs*

[13] O'Connor (2004), p. 71.

[14] Op ci 2, para. 28.

[15] Calabrese (2024), p. 693.

[16] Guillem Carrau (2021), p. 709.

[17] Kur (2023), p. 87.

system legal framework, namely (in brief) where there is some link between the characteristics of the product and its geographical provenance'.[18]

The subject-matter of the dispute was the word 'Warsteiner', the adjectival form of 'Warstein' as a place name, used on the labels on beer bottles. The reason of the dispute was that the beer was brewed in another area, Paderborn. The design of the labels was considered misleading. AG Opinion considered that it was common ground that simple geographical indications of source, in the case of which there was no link between the characteristics of the product and its geographical provenance, did not fall within the definition of the EU GIs system and were not therefore protected.

The Court considered that the EU GIs system (at that time, according to Article 2(2)(b), Regulation No 2081/92), it only concerns GIs in respect of which there is a direct link between both a specific quality, reputation or other characteristic of the product and its specific geographical origin.[19]

There is nothing in the EU GIs system to indicate that such geographical indications of source cannot be protected under the national legislation of a Member State when there is not such a direct link between the geographical origin of the product and its quality, the reputation and any other characteristic of the product. On the top of this, as said in *Haus Cramer* case, the EU GIs system does not preclude application of national legislation which prohibits the misleading use of a simple geographical indication of source, that is, an indication in the case of which there is no link between the characteristics of the product and its geographical provenance.[20]

To conclude, the exclusivity for the Commission on the EU GIs scheme provoked the proliferation of quality labels for agricultural products and foodstuff in some Member States, like the French label rouge. Under certain conditions, Member States allow the use of indications that consist of signs and/or terms explicitly chosen to call consumer attention. They pursued to evoke a characteristic, origin or provenance that make them distinguishable to others[21] but they were not part of the EU GIs system.

3.3 An EU Title via Register: *Salame Felino*

The legal effects of the registration have been fully described in *Salame Felino*,[22] where the Court said that '*the system of protection does not afford protection to a geographical designation which has not obtained a EU registration, but that that geographical designation may be protected, should the case arise, under national legislation concerning geographical designations relating to products for which there is no specific link between their characteristics and their geographical origin,*

[18] AG Opinion, C-312/98—Haus Cramer, ECLI:EU:C:2000:599, para. 43, 44 y 54.

[19] Joined Cases C-321/94 to C-324/94 *Pistre and Others* ECLI:EU:C:1997:229, para. 35.

[20] C-312/98—Haus Cramer, ECLI:EU:C:2000:599, para. 43, 44 y 54.

[21] González Vaqué (1998), p. 33.

[22] C-35/13, Assica and Krafts Foods Italia, ECLI:EU:C:2014:306, para. 27 and 31.

provided, however, that, first, the implementation of that legislation does not undermine the objectives pursued by the EU Regulation (at that time it was Regulation No 2081/92 as amended by Council Regulation (EC) No 535/97) and, secondly, it does not contravene the principle of the free movement of goods under Article 28 EC, matters which fall to be determined by the national court'.

Therefore, the absence of a registration at EU level delivers into a situation where there is not protection of the sign but the unfair competition law or a national system of protection not linked to the geographical origin of the product.

Only the basis of a single European title grants the free movements of goods and prevents the creation of indications of provenance in the different Member States and the problems of the genericity.[23]

The incorporation to the EU Registry is compulsory and constitutive. It has legal effects for the applicants and for those producers that, established within the limits of the GIs territory, would like to produce accordingly to the specifications and, even though, to participate in the managing of the GI. The production of legal effects from the very moment of the registration is a common point with the trade mark and design schemes.[24] In other words, it is a full and completed right. It births only from the administrative act of the registration in a sense that the effectivity is deployed after it.[25]

As defined in *Epoisses* the provisional national protection will not survive after the registry at EU level.[26] Likewise, in *Port Charlotte*, the Court clearly says that the EU protection cannot be completed at national level with an additional protection.[27]

3.4 *Budějovický Budvar Beer*: An Exclusive and Exhaustive in Nature EU System that Prevents Member States from Running a Separate, Parallel National or Regional Scheme

In *Budějovický Budvar*, it was undoubtedly stated that the EU system is exclusive and prevents Member States from running a separate, parallel national or regional scheme (as in case with trademarks, for example).[28]

The reference was made in proceedings between Budějovický Budvar, a brewery established in the town of Česke Budějovice (Czech Republic), and Rudolf Ammersin GmbH ('Ammersin'), a firm established in Vienna (Austria) which distributed beverages, with regard to Budvar's application to prohibit Ammersin from marketing under the mark 'American Bud' beer produced by the brewery Anheuser-Busch Inc. ('Anheuser-Busch'), established in Saint Louis (United States), on the

[23] Beier and Knaak (1994), pp. 30 and 37.

[24] Lobato García-Miján (1995), p. 1987.

[25] Gondra Romero (1996), p. 846.

[26] C-129/97 y C-130/97, Chiciak y Fol, EU:C:1998:274, para. 25–26.

[27] C 56/16 P, PORT CHARLOTTE, EU:C:2017:693, para. 107–108.

[28] C-478/07 Budějovický Budvar, ECLI:EU:C:2009:521, para. 114.

ground that, by virtue of bilateral agreements between the Czech Republic and the Republic of Austria, the use of the designation 'Bud' in that Member State was reserved for beer produced in the Czech Republic.[29]

In the case at hand, the Court concluded that the aim of the EU GIs system was not to establish, longside national rules which may continue to exist, an additional system of protection for qualified geographical indications, like, for example, that introduced by Council Regulation (EC) No 40/94 of 20 December 1993 on the Community trade mark[30] but to provide a uniform and exhaustive system of protection for such indications.

The EU GIs system is exhaustive in nature, with the result that that legislation precludes the application of a system of protection laid down by agreements between two Member States, such as the bilateral instruments at *Budvar,* which conferred on a designation, recognised under the law of a Member State as constituting a designation of origin, protection in another Member State where that protection was actually claimed despite the fact that no application for registration of that designation of origin had been made in accordance with the EU GIs system.

As a matter of fact, the Court described (see para. 116–129 of the Judgment) a few characteristics of the system of protection and supported the view that the system is exhaustive in nature. Unlike other European systems for the protection of industrial and commercial property rights, such as those of the EUTM, the EU registration procedure for GIs is based on powers shared between the Member State concerned and the Commission, since the decision to register a designation may be taken by the Commission only if the Member State concerned has submitted to it an application for that purpose and such an application may be made only if the Member State has checked that it is justified.

The national registration procedures are therefore incorporated in the EU decision-making procedure and constitute an essential part thereof. They cannot exist outside the EU system of protection. Regarding the EU registration procedure, it is also indicative that it provides that, on its sole responsibility, a Member State may grant transitional national protection until a decision on the application for registration is adopted. In that regard, the Court has held, furthermore, that it follows from that provision that, under the system introduced by Regulation No 2081/92, where

[29] The case concerned the interpretation of former Articles 28 EC and 30 EC, of the Act concerning the conditions of accession of the Czech Republic, the Republic of Estonia, the Republic of Cyprus, the Republic of Latvia, the Republic of Lithuania, the Republic of Hungary, the Republic of Malta, the Republic of Poland, the Republic of Slovenia and the Slovak Republic and the adjustments to the Treaties on which the European Union is founded (OJ 2003 L 236, p. 33; 'the Act of Accession'), of Commission Regulation (EC) No 918/2004 of 29 April 2004 introducing transitional arrangements for the protection of geographical indications and designations of origin for agricultural products and foodstuffs in connection with the accession of the Czech Republic, Estonia, Cyprus, Latvia, Lithuania, Hungary, Malta, Poland, Slovenia and Slovakia (OJ 2004 L 163, p. 88) and of Council Regulation (EC) No 510/2006 of 20 March 2006 on the protection of geographical indications and designations of origin for agricultural products and foodstuffs (OJ 2006 L 93, p. 12).

[30] OJ 1994 L 11, p. 1.

Member States have the power to adopt decisions, even of a provisional nature, which derogate from the provisions of the regulation, that power is derived from expressed rules. As the Advocate General observed in point 102 of his Opinion in *Budvar*, a provision of that kind would be meaningless if the Member States were able in any event to retain their own systems of GIs within the meaning of Regulations No 2081/92 and No 510/2006 and have them coexist with those regulations.

Likewise, the fact that EU system is exhaustive in nature was also evidenced by the transitional arrangements for existing national designations such as the Czech designation 'Bud' at issue in the referred proceedings. Thus, Article 17 of Regulation No 2081/92 introduced a so-called 'simplified' registration procedure for national designations legally protected in the Member State which sought their registration or, in Member States where there was no system of protection, validated through use.[31] It was provided that, if their registration was sought within 6 months, their national protection could, on certain conditions, be maintained for a transitional period of 5 years in accordance with Article 13(2) of that regulation, which period was subsequently extended by an additional period of 5 years by Regulation No 535/97.

Since that specific system for the transitional protection of existing national designations had been repealed by Article 1(15) of Regulation No 692/2003, the Commission, by Regulation No 918/2004, laid down transitional provisions regarding the protection of designations of origin and geographical indications existing in the 10 States which became Member States of the European Union at the time of its 2004 enlargement, based on those laid down for the 15 old Member States. Those provisions were especially relevant to the *Budvar* case since, according to the national court, the designation 'Bud' existed at that time as a geographical designation in the Czech Republic. Regulation (EC) No 918/2004 provided that the national protection of geographical indications and designations of origin within the meaning of Regulation (EEC) No 2081/92 existing in the 10 new Member States on 30 April 2004 could be maintained for 6 months. However, as the simplified procedure no longer existed, that regulation provided that where an application for registration had been sent within that 6-month period, that national protection could be maintained until the Commission adopted a decision in accordance with.

As quoted before, in *Port Charlotte,* the European Court of Justice referred that the precise conditions and scope of that protection are laid down exclusively in the EU Law, which governs, in a uniform and exclusive manner, both the authorisation of and limits to, and even the prohibition of, commercial use of PDO and PGI so that, in that specific context, there was no need to apply the conditions for protection specifically established in the relevant rules of Portuguese law which were the basis for the entry of the appellations of origin "Porto" or "Port" in the E-Bacchus database.[32]

---

[31] C-66/00 Bigi, ECLI:EU:C:2002:397, para. 28.

[32] Op cit. 27, para. 96.

In *Champanillo*,[33] the Court recalled that the protection scheme laid down in the, at that time, *lex specialis* applicable to wine was exhaustive as follows:

> 27 However, with regard to the application of Regulation No 1234/2007, the Court has held that the system of protection of PDOs is uniform and exhaustive in nature, with the result that that regulation precludes both the application of a national system of protection of designations or geographical indications and the application of a system of protection laid down by agreements between two Member States which confers on a designation, recognised under the law of a Member State as constituting a designation of origin, protection in another Member State (see, to that effect, judgments of 8 September 2009, Budějovický Budvar, C-478/07, EU:C:2009:521, paragraphs 114 and 129, and of 14 September 2017, EUIPO v Instituto dos Vinhos do Douro e do Porto, C-56/16 P, EU:C:2017:693, paragraphs 100 to 103).

Before examining the questions referred for a preliminary ruling, in *Champanillo*, the CJEU recalled again that GI protection was uniform and exclusive and, as such, precluded both the application of a national system of protection for GIs that were protected under that law and a system of protection laid down by agreements between two Member States. Hence, since the 'Champagne' PDO is protected under the Wine Regulation, neither the bilateral Convention between the French Republic and the Spanish State of 27 June 1973 on the protection of designations of origin, indications of provenance and names of certain products, nor the French Decree No 2010-1441 of 22 November 2010 on the protected controlled designation of origin 'Champagne' and the French Rural Code, were applicable to the dispute. Only the EU Wine Regulation, applicable at the time of the case at hand, was relevant.[34]

3.5 The Protection of the Origin and Its Qualities and Characteristics: *Sekt and Weinbrand*

When addressing the different features of the case *Windsurfing Chiemsee*,[35] the Court defined the characteristics of a geographical indication of origin to become a trade mark in the context of the Trade mark Directive.[36] Therefore, Article 3(3) of the Directive provided that a sign might, through use, acquire a distinctive character which it initially lacked and thus be registered as a trade mark. It is therefore through the use made of it that the distinctive character which is a prerequisite for its registration. In the case at hand, the Court highlighted the specific capacity of the signs or indications that could serve as an indication of the geographical origin:

[33] C-783/19, Comité Interprofessionnel du Vin de Champagne v GB ECLI:EU:C:2021:713, para. 27.

[34] Vrins (2021), p. 1171.

[35] C-108/97—Windsurfing Chiemsee, ECLI:EU:C:1999:230, para. 16.

[36] First Council Directive 89/104/EEC of 21 December 1988 to approximate the laws of the Member States relating to trade marks (OJ 1989 L 40, p. 1).

26. As regards, more particularly, signs or indications which may serve to designate the geographical origin of the categories of goods in relation to which registration of the mark is applied for, especially geographical names, it is in the public interest that they remain available, not least because they may be an indication of the quality and other characteristics of the categories of goods concerned, and may also, in various ways, influence consumer tastes by, for instance, associating the goods with a place that may give rise to a favourable response.

The Judgment delivered in *Sekt and Weinbrand*,[37] can hardly be bettered as regards the relevance of the protection of the geographical origin and its impact in the qualities and characteristics as one of the pillars of the EU GIs system. Since it originated in a specific geographical area, the origin must confer on it a specific quality and specificities of such a nature as to distinguish it from all other products:

7. Whatever the factors which may distinguish them, the registered designations of origin and indirect indications of origin referred to in that directive always describe at the least a product coming from a specific geographical area. To the extent to which these appellations are protected by law they must satisfy the objectives of such protection, in particular the need to ensure not only that the interests of the producers concerned are safeguarded against unfair competition, but also that consumers are protected against information which may mislead them. These appellations only fulfil their specific purpose if the product which they describe do in fact possess qualities and characteristics which are due to the fact that it originated in a specific geographical area. As regards indications of origin in particular, the geographical area of origin of a product must confer on it a specific quality and specific characteristics of such a nature as to distinguish it from all other products.

In the case at hand, the Court stated that Germany failed to fulfil its obligations under the EEC Treaty, as regards the prohibition on measures having an effect equivalent to quantitative restrictions on imports, by reserving the appellations 'Sekt' and 'Weinbrand' to the domestic product and the appellation 'Prädikatssekt' to wines produced within the country from a fixed minimum proportion of home-grown grapes.

In *Ravil Bellon*[38] the question of the extent of the protection afforded by industrial property in the form of PDO was raised as regards the PDO *Grana Padano* and its use only if the cheese is also grated and packaged in the region of production. The Court asserted in a concise paragraph their understanding of this intellectual property right connected to values associated to geographical origin, special characteristics, reputation and authenticity as follows:

49. Designations of origin fall within the scope of industrial and commercial property rights. The applicable rules protect those entitled to use them against improper use of those designations by third parties seeking to profit from the reputation which they have acquired. They are intended to guarantee that the product bearing them comes from a specified geographical area and displays certain particular characteristics. They may enjoy a high reputation amongst consumers and constitute for producers who fulfil the conditions for using them an essential means of attracting custom. The reputation of designations of origin

---

[37] C-12/74—Commission v Germany, ECLI:EU:C:1975:23, para. 7.

[38] C-469/00, Ravil, ECLI:EU:C:2003:295, para. 49.

depends on their image in the minds of consumers. That image in turn depends essentially on particular characteristics and more generally on the quality of the product. It is on the latter, ultimately, that the product's reputation is based (see Belgium v Spain, paragraphs 54 to 56). For consumers, the link between the reputation of the producers and the quality of the products also depends on his being assured that products sold under the designation are authentic.

As said by the Court the justification of protecting a GI based on its reputation was to avoid an improper use of the term by other stakeholders that were seeking to profit from the mentioned GIs reputation. The Court linked GIs producers' reputation and the quality of the product by the image of the GIs' authenticity in the mind of consumer.

To sum up, from the referred case law and as it is a very well known fact, it can be concluded that the EU GIs system relies on the fact that qualities and characteristics are due to the specific geographical area of origin.

3.6 The Protection of an Intrinsic Reputation: *Nero Champagne*

As remarked before, PGIs can be registered only because of its reputation. This is a critical issue for the EU GIs scheme. For instance, in the *Feta* case, the link between '*the name "Feta" and Greece is thus deliberately suggested and sought as part of a sales strategy that capitalises on the reputation of the original product, and this creates a real risk of consumer confusion*'.[39]

In *Bavaria*,[40] the Court said that '*...the registration of a PGI under Regulation No 2081/92 is designed, among other objectives, to prevent the improper use of a name by third parties seeking to profit from the reputation which it has acquired and, moreover, to prevent the disappearance of that reputation as a result of popularisation through general use outside its geographical origin or detached from a specific quality, reputation or other characteristic which is attributable to that origin and justifies registration*'.

In *Colombueno*,[41] the Board of Appeal recalled the EU legislator stated that GIs can be registered because of their reputation. This reputation was argued by the producer groups when applying for a recognition of the GI.[42] Therefore, GIs are intrinsically recognised merely because they are registered. This is independent of whether a GI has been registered based on a claim of reputation essentially attributable to its geographical origin. The decision was not appealed.

[39] Joined Cases C-465/02 and C-466/02, Denmark v Commission, Germany v Commission, ECLI:EU:C:2005:804 and ECLI:EU:C:2005:636, para. 20.

[40] C 343/07, Bavaria and Bavaria Italia, EU:C:2009:415, para. 106.

[41] 01/10/2021, R 566/2015-2, COLOMBUENO / CAFÉ DE COLOMBIA, para. 67.

[42] Article 9 Regulation (EU) 2024/1143.

In *Nero Champagne*[43] EUIPO Boards of Appeal's decision, it was discussed if the PDOs enjoy or not an intrinsic reputation by the mere fact that they are registered, given that the legislator specifically envisaged one scenario, which required reputation (Article 103(2)(a)(ii) Regulation (EU) No 1308/2013). The PDO intrinsic reputation was sustained by the EUIPO Opposition Division but not shared by the Board of Appeal as they concluded that, in any event, it was proven that the PDO 'Champagne' enjoys an exceptional reputation in the EU.

> 30 According to the Opposition Division, since all registered PDOs offer a guarantee of quality deriving from their geographical origin, PDOs are generally considered to enjoy an intrinsic reputation, within the meaning of Article 103(2)(a)(ii) of Regulation (EU) No 1308/2013, by the mere fact that they are registered. The Board is not convinced by this interpretation given that in Article 103(2)(a) of Regulation (EU) No 1308/2013, the legislator specifically envisaged two different scenarios, only one of which requires reputation. A different interpretation would deprive Article 103(2)(a)(ii) of any effet utile.
>
> 31 In any event, the opponents filed evidence of reputation and have shown that the PDO 'Champagne' enjoys an exceptional reputation in the EU and conveys an image of excellence, prestige and celebration, as also confirmed by the case-law: see, among others, District Court of Paris, Third Chamber, First section, 05/10/1984, A.R.7147-9227-87., CIVC - Pol Roger v. SEITA (Annex VII.3); Court of Appeal of Caen, 28/10/2003, 02/2403, CIVC v. Hotel-Restaurant "Le Champagne" (Annex VII.4.); 17/04/2020, R 1132/2019-4, Champagnola / Champagne, § 42; Audiencia Provincial de Barcelona, Civil, 18/03/2022, 1726/2018, Champanillo, § 10.4. Furthermore, the applicant does not challenge that the PDO 'Champagne' is reputed in the EU.

Therefore, the EUIPO in its first instance decision considered that GIs are intrinsically reputed by the mere fact that they are registered[44] but the Board of Appeal considered that the opponents filed evidence of reputation and showed that the PDO 'Champagne' enjoys an exceptional reputation in the EU and conveys an image of excellence, prestige and celebration.[45] As it is known, *Nero Champagne* Decision was partially annulled by the General Court. In *Nero Champagne* Judgment, the Court noted that EU law a trade mark can contain a PDO. However, its registration may be refused if the PDO does not comply with the product specification con-

[43] 17/02/2023, R 531/2022-2, NERO CHAMPAGNE / Champagne (GI).

[44] Therefore, the Office considers that GIs are intrinsically reputed within the meaning of Article 26(1)(a) of Regulation (EU) 2024/1143 by the mere fact that they are registered. This is irrespective of whether a GI has been registered on the basis of a claim in the application to its reputation being essentially attributable to its geographical origin (Article 46(1)(b) and Article 46(2)(b) of Regulation (EU) 2024/1143, for agricultural products, Article 93(1)(b)(i) of Regulation (EU) No 1308/2013, for wines, and Article 3(4) of Regulation (EU) 2019/787, for spirit drinks). See EUIPO Guidelines (2024) Part C, Section 6.3.1.1 Use including exploitation of the GI.

[45] The Board of Appeal (see para. 31 of the Decision) added that the reputation is also confirmed by the case-law: see, among others, District Court of Paris, Third Chamber, First section, 05/10/1984, A.R.7147-9227-87., CIVC—Pol Roger v. SEITA (Annex VII.3); Court of Appeal of Caen, 28/10/2003, 02/2403, CIVC v. Hotel-Restaurant "Le Champagne" (Annex VII.4.); 17/04/2020, R 1132/2019-4, Champagnola / Champagne, § 42; Audiencia Provincial de Barcelona, Civil, 18/03/2022, 1726/2018, Champanillo, para. 10.4.

cerned, if its use exploits the reputation of a PDO or if the mark applied for conveys a false or misleading indication as to the provenance or origin of the product (para. 31). Any trade mark is liable to exploit unduly the reputation of a PDO, even if it designates only products which comply with the product specification of that PDO. Thus, when evidence to that effect is brought to EUIPO's attention, EUIPO must examine it in order to ascertain whether they allow that presumption to be rebutted (para. 54) and, if it is considered that the restriction is supported by a presumption *iuris tantum*, EUIPO (including BoA) should have set out the reasons to a sufficient legal standard of why the reputation is not overturned (para. 66).[46]

Therefore, there is a second and different idea of reputation, when it is said that the GI is protected against undue exploitation of their reputation (Articles 26 and 31 Regulation (EU) 2024/1143). In this second scenario, EUIPO Guidelines describe that the opponents (the GIs) do not have to submit evidence of the reputation of the GI. However, they must submit convincing arguments and/or evidence regarding the exploitation of the reputation of the GI. As a rule, general allegations (such as merely citing the relevant wording of the EU regulations) of exploitation of the reputation will not be sufficient in themselves to prove such exploitation. The opponent must adduce evidence and/or develop a cogent line of argument to demonstrate specifically how the alleged injury might occur, considering both the rights concerned, the goods and services in question and all the relevant circumstances.[47]

Likewise, an academic debate has been generated around the fact that PGIs can be protected based on its reputation.

When describing the PGI, it is only name, hence only the designations are directly protected and not the traditional methods of production; it requires only one of the production steps to take place in the defined area; and the simple "reputation" of the product can be the base of the link with the geographical area. This link, however, cannot be scientifically proved unlike the physical/environmental link between the product and its place of origin required for PDOs ("terroir products"). On the other hand, Regulation (EU) 2024/1143 clearly acknowledges the importance of protecting traditional goods and fostering rural development (see recitals 2 and 4 of the Preamble). However, the text is sufficiently broad to accommodate different kinds of producers, and this gap cannot be deemed unintentional. In the context of the GIs, the meaning to be given to the "reputational" requirement is of paramount importance. Indeed, this concept is still particularly under-researched. It is known that the link can be very weak. Furthermore, an analysis of the specifications of the registered PGIs shows that the link is often proved through (1) historical recounts; (2) literary references; (3) festivals and fairs; (4) tales; (5) newspaper

---

[46] T-239/23, Comité interprofessionnel du vin de Champagne and INAO v EUIPO—Nero Lifestyle (NERO CHAMPAGNE) ECLI:EU:T:2025:638.

[47] See EUIPO Guidelines (2026) Part C, Section 6.3.1.1 Use including exploitation of the reputation of the GI.

archives; (6) other.[48] However, it is necessary to stress that the "reputational" link does not necessarily influence the content of the specification.[49]

PGIs and PDOs are complex rights—more complex than trade marks (only collective marks and certification marks have some proximity—even so, there are several differences). PGIs and PDOs fulfil three legal functions: geographical origin; quality; and publicity; being the geographical origin one softer for PGIs, as currently only one of the production steps needs to take place inside the defined geographical area or in the case of a wine GI that 15% of the grapes used in its production come from outside. As regards quality, on PGI, the quality guarantee function may only be based on reputation. In fact, even if it is true that the product comes entirely from a certain region or place (which name corresponds to the PGI) we are facing here not a PGI, but rather a simple indication of source that has a reputation. Was the aim of the legislator to qualify as a PGI a simple indication of source that has (and only has) reputation (no other link with the region)? Ribeiro considers that, however, to guarantee the integrity of the legal system and the coherence of the industrial property right, this kind of PGI (based only on reputation) should not have this qualification (an industrial property right) and be only protected through the unfair competition rules.[50]

Therefore, firstly, the reputation of a PGI can justified its registration. Secondly, the extend of the protection under EU GIs scheme spread to any type of undue exploitation weakening, dilution or detriment to the reputation or evocation of their reputation. Articles 26 and 31 Regulation (EU) 2024/1143 no longer limit the EUIPO ex officio examination to identical and comparable goods. These situations can result from the information available in the Office or following third party observations. In ex parte cases, EUIPO's practice consists of not asking to the opponents (the GIs) the submission of evidence of the reputation of the GI but of arguments in favour of the existence of the exploitation of the reputation.

3.7 The *Manchego*: Only for All the Eligible Producers that Respect the Product Specification

Producer groups play an essential role in the application process for the registration and in the management of the GIs. The protection of the origin expands its cover in the territory of origin by affecting not only producers under the GI but others. The *Manchego cheese* case solves a preliminary ruling request in proceedings concerning, inter alia, the use by on stakeholder of labels to identify and market cheeses which were not covered by the PDO 'queso manchego'.

Industrial Quesera Cuquerella SL marketed certain cheeses using labels with an illustration of a knight similar to the usual depictions of Don Quixote de La Mancha, a bony horse and landscapes with windmills and sheep, as well as the words 'Quesos

[48] Gangjee (2012), p. 222.

[49] Zappalaglio (2015).

[50] Ribeiro de Almeida (2021), p. 302.

Rocinante' ('Rocinante cheeses'). Those images and the word 'Rocinante' refer to the novel Don Quixote de La Mancha by Miguel de Cervantes, 'Rocinante' being the name of the horse ridden by Don Quixote. The cheeses in question were not covered by the PDO 'queso manchego', which covers cheeses made in the region of La Mancha (Spain) from sheep's milk in accordance with the requirements of the product specification of that PDO.

In its judgment, the Court held that a registered name might be evoked by figurative signs, including where such figurative signs were used by a producer established in that region, but whose products, similar or comparable to those protected by the designation of origin, were not covered by the PDO. However, the wording of Article 13(1)(b) of Regulation No 510/2006 (now Article 26 (1) of Regulation (EU) 2024/1143) did not provide that a producer established in a geographical area corresponding to the PDO and whose products are not protected by the PDO but are similar or comparable to those protected by it is to be excluded from that provision, concluded the Court. Consequently, in a situation such as that at issue in the *Manchego* proceedings, the fact that a producer of similar or comparable products to those protected by a designation of origin is established in a geographical area associated with that designation cannot exclude that producer from the scope of the GI protection.[51]

In fact, AG Opinion considered in the case at hand,[52] allowing producers which operate in the geographical area associated with a given GI, but do not comply with the relevant requirements, to use signs evoking that geographical area for products identical or similar to those covered by the GI would weaken the assurance of quality conferred by that name, a guarantee that, although essentially related to the geographical origin of GI products, requires compliance with all the provisions governing the matter. Moreover, not to extend the protection from evocation to conduct by local producers that do not comply with the requirements would undermine the rights that must be reserved for producers who have made a genuine effort to improve quality to be able to use a GI.

Therefore, the EU GIs System relies on the recognition of the exclusive use of the protected term to all the eligible producers that respect the product specification. The ultimate objective of the legislation on PDOs and PGIs is therefore to protect traditional products 'with specific characteristics linked to geographical origin'. Recital 18 of Regulation (EU) 2024/1143 declares that GIs are a collective right held by all eligible producers in a designated area willing to adhere to a product specification. This last expression contained in a recital is a colossal novelty because till this last reform the legislator had not clarified the doubts about the ownership of the GI.[53]

[51] C-614/17, Fundación Consejo Regulador de la Denominación de Origen Protegida Queso Manchego, ECLI:EU:C:2019:344, para. 34 and 36.

[52] C-614/17, Fundación Consejo Regulador de la Denominación de Origen Protegida Queso Manchego, ECLI:EU:C:2019:11, para. 43.

[53] Maroño Gargallo (2002), p. 261.

3.8 *Grana Biraghi:* The Legislative Position and Its Historical Development as Regards the Protection of a GI

From *Grana Biraghi*, we can extract some conclusions about the relevance of the legislative *state of arts* and its historical precedents on the protection of a GI. In the case at hand, the EUIPO Boards of Appeal concluded that the existence of the PDO 'grana padano' did not preclude the registration of the sign GRANA BIRAGHI as a *Community trade mark*. The decision was brought to the Court by the *Consorzio per la tutela del formaggio Grana Padano* and the Court concluded that Boards of Appeal erred in finding that the name 'grana' was generic.[54]

In its judgment, the Court showed the importance of duly take into consideration all the evidence provided by the applicant and the criteria identified by the case-law of the Court to conclude that proof of the generic character of the name 'grana' had not been adduced to the requisite legal standard. In a detailed reasoning, the Court referred to the legislative position as regards the protection of the name 'grana padano' in Italy and the historical development thereof:

> 74. In that regard, the Court notes that the first legislative recognition of the name 'grana' dates back to Regio Decreto Legge No 1177, Disposizioni integrative della disciplina della produzione e della vendita dei formaggi (Royal Decree-Law No 1177 of 17 May 1938 introducing provisions to supplement the rules on the production and sale of cheeses, GURI No 179 of 8 August 1938). In that decree, which lays down the minimum fat content in the different Italian cheeses, reference is made to grana parmigiano-reggiano, grana lodigiano, grana emiliano, grana lombardo and grana veneto. The Decree shows that grana was produced in a number of areas in the Padanian Valley, near Parma, Reggio Emilia and Lodi, and in Emilia, Lombardy and Venetia. The name 'grana padano', by contrast, is not mentioned in the Decree.
>
> ....
>
> 76 Following the introduction of the first system of designations of origin in Italy by Law No 125/54 and the Decreto del Presidente della Repubblica No 1269, Riconoscimento delle denominazioni circa i metodi di lavorazione, caratteristiche merceologiche e zone di produzione dei formaggi (Decree of the President of the Republic No 1269 of 30 October 1955 on the recognition of designations relating to methods of production, characteristics of marketing and areas of production of cheeses, GURI No 295, 22 December 1955, p. 4401) which recognised the designation of origin 'grana padano', parmigiano reggiano cheese lost its classification as grana, on account of its specific characteristics, whilst all the other grana cheeses were recognised under the designation 'padano'.

In particular, the Court was concerned about the fact of the Italian legislation referring to different *granas* without mentioning *grana padano* and the posterior insertion of the qualifier 'padano' not to restrict the scope of the PDO to certain *granas* only, but to place them all under the same increased protection:

> 77 The fact that the Italian legislation of 1938 refers to different granas (parmigiano-reggiano, lodigiano, emiliano, lombardo and veneto), all produced in the area of the plain

[54] T-291/03, Grana Biraghi, EU:T:2007:255, para. 74–78.

of the Po, without however mentioning grana padano, and also the fact that subsequent legislation introduced the name 'grana padano' whilst abandoning the earlier names indicates that grana is a cheese traditionally produced in numerous areas of the plain of the Po, which the Italian legislature therefore, at a certain point in time, identified by the term 'padano' so as to simplify the framework of rules and to include in one single name the various previous names, all originating in the Padanian Valley.

78 The qualifier 'padano' was therefore inserted not to restrict the scope of the PDO to certain granas only, but to place them all under the same increased protection, conferred initially by Italian legislation and subsequently by Regulation No 2081/92. It follows from this that the changes in the Italian legal context indicate that the name 'grana' is not generic.

None of the arguments submitted by the trade mark applicant made it possible to call the referred findings into question; for instance, the Decisions of the European Commission referring to the Danish cheese 'Dansk grana' and the German cheese 'Rosmonte-Typ grana' in the context of Council Directive 92/46/EEC; the frequency with which a term appears on the internet; the fact that the term 'grana' does not designate a geographical area as such or that the bulletin of the association Assolatte showed that there were 'other granas' in addition to grana Padano. For the Court, they were not, of themself, capable of establishing the generic nature of a name. In *Grana Biraghi,* it is shown how GIs protection depends on the legislative protection and its historical development and not on definitions of the dictionaries or internet extracts.

3.9 *Aceto Balsamico di Modena*: The Exclusive Protection of the Protected Term As Such

Generic terms cannot be registered in any case or at any moment because the essence of the generic term is the absence of distinctiveness and their presence in the Registry demerits the protected GIs.[55]

In *Bavaria,*[56] the Court stated that '*as regards a PGI, a name becomes generic only if the direct link between, on the one hand, the geographical origin of the product and, on the other hand, a specific quality of that product, its reputation or another characteristic of the product, attributable to that origin, has disappeared, and that the name does no more than describe a style or type of product*'.

However, a GI cannot become generic. It is for the Commission to determine whether a name is generic or not, and it will do so in accordance with the procedure specifically defined by the Regulation, after canvassing informed opinion and considering all the evidence supporting both sides of the argument. Since the assessment of whether a name is generic falls within the remit of the Commission,[57] it is

[55] Guillem Carrau (2021), p. 320.

[56] C-343/07, Bayerisches Bier, EU:C:2009:415, para. 107.

[57] And, to a lesser extent, of the Council, because under Article 3(3) of the regulation, before the entry into force of the regulation, the Council, acting on a proposal from the Commission, is to draw up a list of the names of the agricultural products or foodstuffs which are regarded as being generic.

not for the Court to take on the task of the Commission on that question. The role of the Court consists simply in reviewing the legality of decisions adopted by the Commission (or the Council) on the subject[58]

A clear example of this situation is the *Aceto Balsamico di Modena* case where it raises the issue of whether the word 'balsamico' is entitled to protection as a GI. This is the essential question which arises in the request for a preliminary ruling, which was lodged by the Bundesgerichtshof (Federal Court of Justice, Germany). At that moment, the definition of 'generic terms' contained in Article 3(6) of Regulation No 1151/2012 was very specific and limited in scope: '*the names of products which, although relating to the place, region or country where the product was originally produced or marketed, have become the common name of a product in the Union*'. The definition thus referred to terms which over time have lost their geographical connotation and were merely a common name.

As the AG Opinion highlighted in the case at hand, the registration of the name 'Aceto balsamico tradizionale di Modena (PDO)' pursuant to Regulation No 813/2000, which is all but identical to the PGI 'Aceto Balsamico di Modena', save for the additional word 'tradizionale' and the capitalisation of the 'b' in 'balsamico' clearly argued, in favour of a finding that only the indication 'Aceto Balsamico di Modena' as a whole is protected and that the terms 'Aceto', 'balsamico' and 'Aceto balsamico' are simply common words. Such an approach was unequivocally supported by recitals 8 and 9 of Regulation No 583/2009.[59]

In *Aceto Balsamico di Modena,* the General Court refers that protection covers the whole term and it does not extend to the use of the non-geographical terms of that name such as 'aceto' and 'balsamico'. The exclusive use is limited to the protected term and not to the parts of the term that are not expressly protected.[60]

3.10 Sustainability: The New Component Defined by the Legislator

GIs can play an important role in terms of sustainability, including in the circular economy, thereby enhancing their heritage value and thus strengthening their role within the framework of national and regional policies with a view to meeting the objectives of the European Green Deal, as the legislator says in recital number 2 of Regulation (EU) 2024/1143.

The Commission communication of 20 May 2020 entitled 'A Farm to Fork Strategy — for a fair, healthy and environmentally-friendly food system', which called for a transition to sustainable food systems, also called for the strengthening of the legislative framework on GIs and for the inclusion of specific sustainability criteria. In that communication, the Commission committed to strengthening,

[58] See AG Opinion in C-66/00, op. cit. 24, para. 39–40.

[59] See AG Opinion in C-432/18—Consorzio Tutela Aceto Balsamico di Modena. ECLI:EU:C:2019:1045, para. 57.

[60] C-432/18—Consorzio Tutela Aceto Balsamico di Modena. ECLI:EU:C:2019:1045, para. 35–36.

among other players, the position of producers of products with GIs, their cooperatives and producer organisations in the food supply chain. Focus should be placed on small-scale producers, particularly those who best preserve traditional skills and know-how.[61]

This purpose is defined in Article 7 Regulation (EU) 2024/1143 where it is said that a producer group may agree on sustainable practices to be adhered to in the production of the product designated by a GI or in carrying out other activities subject to one or more obligations provided for in the product specification.

Consistent with EU and international developments, sustainability must be a horizontal principle that underpins the attainment of ambitious new targets for sustainable development. However, the inclusion of additional requirements must be carefully done to facilitate seamless access to the system.[62]

A definition of 'sustainable practice' is included in Article 7 Regulation (EU) 2024/1143. It shall be a practice which contributes to '*…one or more social, environmental or economic objectives, such as:*

(a) climate change mitigation and adaptation, the sustainable use and protection of landscapes, water and soil, the transition to a circular economy, including the reduction of food waste, pollution prevention and control, and the protection and restoration of biodiversity and ecosystems;
(b) the production of agricultural products in ways that reduce the use of pesticides and manage risks resulting from such use, or that reduce the danger of antimicrobial resistance in agricultural production;
(c) animal welfare;
(d) a fair income for producers, diversification of activities, promotion of local agricultural production, and valorisation of the rural fabric and local development;
(e) preservation of agricultural employment by attracting and sustaining young producers and new producers of products benefiting from a geographical indication;
(f) improving working and safety conditions in agricultural and processing activities.'

In this sense, such practices, firstly, must be included in the product specification in accordance with the registration or amendment procedure; and, secondly, must apply sustainability standards higher than those laid down by Union or national law in terms of environmental, social or economic sustainability or animal welfare.

In addition, the legislator considers that the producer group is responsible for drafting and regularly updating a 'sustainability report' that the European Commission will publish. This report must be based on verifiable information, comprising a description of existing sustainable practices implemented in the production of the product, a description of how the method of obtaining the product impacts on sustainability, in terms of environmental, social, economic or animal welfare commitments, and information necessary to understand how sustainability affects the development, performance and position of the product.

---

[61] Recital 4 Regulation (EU) 2024/1143.

[62] Montero García-Noblejas (2022), p. 522.

3.11 Conclusion

From the outset, the EU GIs scheme was defined as a legal solution to different circumstances and social and economic components that were present in the context of the production of high-quality foodstuff products. Some of the key components of the scheme, that have been described in this chapter, were identified by the legislator from the very beginning of the system with the adoption of Regulation (EEC) num. 2081/1992 (for instance, the role of the producers, the requirements for protection, the existence of a register, the basic rules for relations between trade marks and GIs, etc.) others have been confirmed by a consistent case law (for instance, the unitary and exhaustive identity of the system; the producer's role; the 'being a type of intellectual property right'; the ability to use a registered name; the system of controls; the coexistence with homonymous terms; the generic terms, the extension of the protection, etc.).

Consequently, the EU GIs scheme is composed of a set key components, as confirmed through a body of EU CG/CJ jurisprudence, including among others *Bavaria, Rioja II, Turrón de Jijona*, *Sekt*, *Salami felino, Budějovický Budvar, Champagne Nero, Manchego, Grana Biraghi,* and *Aceto Balsamico di Modena.* Our analysis of the case law seeks, like a jigsaw falling into place, to accurately illustrate the GIs' scope of protection, largely rooted in the principles of origin, reputation, exclusivity, non-genericity and sustainability.

References

Beier FK, Knaak J (1994) The protection of the direct and indirect geographical indication of source in Germany and in the European Community. IIC international review of industrial property and copyright law

Calabrese B (2024) The Evolving Protection of Geographical Indications Against Services: "Brand" New World?. IIC 55, p. 348–367.

Gangjee D (2012) Relocating the law of geographical indications. CUP, Cambridge

Gaultier G (2005) La reconquête de la protection des appellations d'origine, Jijona et Alicante (The reconquest of the protection of Jijona and Alicante GIs) in collective work Homenaje a A. Bercovitz, Estudios sobre propiedad industrial e intelectual y Derecho de la competencia, Grupo español de la AIPPI, pp 501–511

Gondra Romero JMª (1996) Teoría General de los signos de empresa (General theory of signs in the commerce) in collective work Estudios en homenaje al Profesor Aurelio Menéndez, coord. Pro Juan Luis Iglesias de Prada, Tomo I, Civitas

González Vaqué L (1998) La Sentencia "Pistre": aplicabilidad del artículo 30 a las denominaciones agroalimentarias protegidas nacionalmente (Pistre case: the applicability of Article 30 to agricultural terms protected at national level). Comunidad Europea Aranzadi 1:33–40

Gragnani M (2012) The law of geographical indications in the EU. J Intellect Prop Law Pract 7(4):271–282

Guillem Carrau J (2021) Denominaciones geográficas de calidad. Estudio de su régimen jurídico en la OMC, la Unión europea y el Derecho español (Quality geographical indications. A study of the legal scheme of the WTO, EU and Spanish law), 2nd edn. Tirant lo blanch

Kur A (2023) No strings attached to GIs? About a blind spot in the (academic) discourse on limitations and fundamental rights. IIC 54:87–94

Lobato García-Miján M (1995) La protección de las denominaciones de origen. Estudio del Reglamento (CEE) núm. 2081/1992 (The protection of designations of origin. Study of Regulation (EEC) n. 2081/1992), en Estudios de Derecho mercantil en homenaje al profesor Manuel Broseta Pont, VVAA, II, Valencia, Tirant lo Blanch, pp 1975 a 1993

Maroño Gargallo MªM (2002) La protección jurídica de las denominaciones de origen en los derechos español y europeo (The legal protection of designations of origin in Spanish and EU Law). Marcial Pons, Madrid

Montero García-Noblejas P (2022) Funciones de las indicaciones geográficas como Derechos de Propiedad Intelectual y Desarrollo Sostenible (GIs functions as intellectual property rights and sustainable development) in Competencia, propiedad intelectual y tutela de consumidores en el sector agroalimentario / coord. por Vanessa Jiménez Serranía; Fernando Carbajo Cascón (dir.), pp 497–523

O'Connor B (2004) The law of geographical indications. Cameron May, London

O'Rourke R (1998) European food law. Palladian Law Publishers, Brussels

Pérez Ferreras S (2004) Quelles sont les différences juridiques qui permetent la distinction entre les indications géographiques et les marques ? (Which are the differences that allow the distinction among GIs and trade marks). AIDV Bull n°34, pp 14–20

Ribeiro de Almeida A (2021) The GI structure (the model behind the EU rules) or the complexity of this subjective right. J Intellect Prop Law Pract 16:4–5

Tinlot R et al (1987) La notion d'apellation d'origine (The notion of designation of origin). Paper presented at Symposium Les apellations d'origine historiques, O.I.V., Jerez

Vrins O (2021) No champagne for 'Champanillo': the protection of PDOs and PGIs against evocation for services. J Intellect Prop Law Pract 16(11):1171–1174

Zappalaglio A (2015) The Piadina Romagnola Mess. A new legal case for an old question: what is a GI? (15 September 2015). https://ssrn.com/abstract=2791400

The Procedural Aspects of Geographical Indications: Registration, Controls and Enforcement

4

Abstract

This chapter is devoted to a set of GIs procedural aspects that are the backbone of the EU GIs system. Based on judgments like the ones on the cases *Port Charlotte*, *Halloumi* and *Comté*, we examine the importance of the registration as an essential condition for protection, the role of Member States in facilitating the reception of the application and in giving a provisional protection, as well as granting the national opposition procedures. *Kühne* and *Jambon sec* cases showed that the general understanding should be that it is for the Member State to check the application; for the Commission to assess that the application contains the information required (i.e. standing orders) and to scrutinise it is not vitiated with errors; and for the National Courts to verify decisions taken by national authorities on minor amendments. We make technical findings on the difficulties that the system faces when the national decision is reviewed by a national Court or when the Member State is called to adopt measures against conduct which infringe a PDO. Furthermore, it deals with the verification and control systems in place to ensure the authenticity and quality of GI products and with the adoption of measures by the authorities against conducts which infringed GIs. Finally, it covers the new provisions from the EU Digital Services Act related to enforcement and online intermediary services as regards the responsibilities of the judicial and administrative authorities as GIs enforcement.

In contrast to other EU-law systems for the protection of industrial and commercial property rights, such as that relating to the EU trade mark, which is provided for by Regulation (EU) 2017/1001 (EUTMR),[1] or that relating to plant variety rights,

[1] Regulation (EU) 2017/1001 of the European Parliament and of the Council of 14 June 2017 on the European Union trade mark (codification) (Text with EEA relevance) *OJ L 154, 16/06/2017*.

© The Author(s), under exclusive license to Springer Nature Switzerland AG 2026

J. Guillem Carrau, *Landmark EU Geographical Indications Case Law*, Law for Professionals, https://doi.org/10.1007/978-3-032-26101-4_4

which is provided for by Regulation (EC) No 2100/94,[2] Regulation (EU) 2024/1143 establishes a single common Union framework for the registration and protection of GIs in three agricultural sectors: wine, spirit drinks and agricultural products.

As it is known, Regulation (EU) 2024/1143 includes exhaustive provisions on the procedures for registration, for the amendment of the product specification and for the cancellation of GIs in all the agricultural sectors (agriculture products, wines and spirit drinks) and traditional specialities guaranteed and, on their protection, and enforcement.[3] It empowers the Commission to adopt delegated and implementing regulation with the aim of completing the legal scheme: Commission Delegated Regulation (EU) 2019/33,[4] Commission Implementing Regulation (EU) 2025/26[5]; and Commission Delegated Regulation (EU) 2025/27[6] provide more detail, and

---

[2] Council Regulation (EC) No 2100/94 of 27 July 1994 on Community plant variety rights, OJ L 227, 01/09/1994.

[3] Regulation (EU) 2024/1143 also includes the rules on controls as regards geographical indications for agricultural products and spirit drinks and traditional specialities guaranteed. In addition, that Regulation contains a specific section concerning geographical indications for the agricultural products sector, providing rules on the definition of geographical indications, content of the product specification and of the single document and specific rules on sourcing of feed and raw materials and on plant varieties and animal breeds. In order to fully harmonise rules on geographical indications in all the agricultural sectors (agriculture products, wines and spirit drinks), Regulation (EU) 2024/1143 also amends certain rules on geographical indications in the wine and spirit drinks sector, which are provided for by Regulations (EU) No 1308/2013 and (EU) 2019/787, respectively. In particular, these provisions concern the definition of protected designations of origin and of protected geographical indications in the wine sector and the definition of geographical indications in the spirit drinks sector, the content of the product specification and of single document in both sectors, the controls in the wine sector and other specific rules.

[4] Commission Delegated Regulation (EU) 2019/33 of 17 October 2018 supplementing Regulation (EU) No 1308/2013 of the European Parliament and of the Council as regards applications for protection of designations of origin, geographical indications and traditional terms in the wine sector, the objection procedure, restrictions of use, amendments to product specifications, cancellation of protection, and labelling and presentation, OJ L 9, 11.1.2019.

[5] Commission Implementing Regulation (EU) 2025/26 of 30 October 2024 laying down rules for the application of Regulation (EU) 2024/1143 of the European Parliament and of the Council as regards registrations, amendments, cancellations, enforcement of the protection, labelling and communication in respect of geographical indications and traditional specialities guaranteed, and amending Implementing Regulation (EU) 2019/34 as regards geographical indications in the wine sector, and repealing Implementing Regulations (EU) No 668/2014 and (EU) 2021/1236, OJ L, 2025/26, 15.1.2025.

[6] Commission Delegated Regulation (EU) 2025/27 of 30 October 2024 supplementing Regulation (EU) 2024/1143 of the European Parliament and of the Council with rules concerning the registration and the protection of geographical indications, traditional specialities guaranteed and optional quality terms and repealing Delegated Regulation (EU) No 664/2014, OJ L, 2025/27, 15.1.2025.

Commission Delegated Regulation (UE) 2025/29[7] repeals some previous rules. The legal framework is completed by Commission Delegated Regulation (UE) 2025/28.[8]

The procedure for registering GIs is based on powers shared between the Member State concerned and the Commission, since the decision to register a designation may be taken by the Commission only if the Member State concerned has submitted to it an application for that purpose and such an application may be made only if the Member State has checked that it is justified.[9]

In cases like *Kühne* and *Bavaria*, the Court has held that the EU quality food scheme created a system of division of powers, in the sense that, in particular, the decision to register a name as a PDO could be made by the Commission only if the Member State concerned had submitted to it an application for that purpose, and that such an application could be made only if that Member State had checked that the application was justified. This system of division of powers is attributable particularly to the fact that registration assumes that it has been verified that a certain number of conditions have been met, which requires largely detailed knowledge of matters particular to that Member State, which the competent authorities thereof are best placed to check.[10] Within this system of division of powers, the legal scheme provides for legal remedies against the unjustified registration of geographical indications and designations of origin both at national and at European level.[11]

4.1 The *Port Charlotte* Case: A Registration Is *conditio sine qua non* for the Protection

The registration constitutes the definitive administrative step towards the definition of the protected term and its contours. In *Port Charlotte,* the Court referred that the precise conditions and scope of that protection are laid down exclusively in the EU Law, which governs, in a uniform and exclusive manner, both the authorisation of and limits to, and even the prohibition of, commercial use of the GIs, so that, in that

[7] Commission Delegated Regulation (EU) 2025/29 of 30 October 2024 repealing Delegated Regulation (EU) 2021/1235 supplementing Regulation (EU) 2019/787 of the European Parliament and of the Council with rules concerning applications for registration of geographical indications of spirit drinks, amendments to product specifications, cancellation of the registration and the register, OJ L, 2025/29, 15.1.2025.

[8] Commission Delegated Regulation (EU) 2025/28 of 30 October 2024 amending Delegated Regulation (EU) 2019/33 supplementing Regulation (EU) No 1308/2013 of the European Parliament and of the Council as regards applications for protection of designations of origin, geographical indications and traditional terms in the wine sector, the objection procedure, restrictions of use, amendments to product specifications, cancellation of protection, and labelling and presentation, *OJ L, 2025/28, 15.1.2025.*

[9] C-56/16 P, Port Charlotte, EU:C:2017:693, para. 86.

[10] See, by analogy, Cases C-269/99, Carl Kühne and Others, EU:C:2001:659, para. 53, and C-343/07, Bavaria and Bavaria Italia, EU:C:2009:415, para. 66.

[11] Günzel (2012), p. 245.

specific context, there was no need to apply the conditions for protection specifically established in the relevant rules of Portuguese law which were the basis for the entry of the appellations of origin "Porto" or "Port" in the E-Bacchus database.[12]

A Member State cannot create any type of registry at national level to keep any variations of the EU protected term applied for. For instance, in a preliminary ruling on the marketing of *Epoisses de Bourgogne* cheese, the Court concluded that a Member State may not, by adopting provisions of national laws, alter a designation of origin for which it has requested registration at EU level and protect that designation at national level.[13]

As confirmed in *Bigi*, this statement goes till the fact that, from the very moment in which the Member States applies for the protection of a PDO under the EU simplified proceedings, the system is not covering goods not meeting the product specification for that PDO.[14]

4.2 The *Halloumi* Cheese and *La Conqueste*: A Set of Tasks for the Member States, Among Others, to Allow Any Stakeholder to Oppose

'Direct applicability' and 'direct effect' are the cornerstone of any EU Regulation.[15] Therefore, it is necessary to describe how important is that EU GIs scheme reflects a list of obligations and tasks for the Member States as part of an administrative procedure envisage to protect a GI at EU level.[16] As it happens with other intellectual property rights, the obtention of the right is subject of a complex procedure that starts with an application for registration and ends with an inscription in a Registry,

[12] Op. cit. 9, para. 96.

[13] C-129/97—Chiciak and Fol ECLI:EU:C:1998:274.

[14] C-66/00—Bigi ECLI:EU:C:2002:397.

[15] Under Art. 288 TFEU 'A Regulation shall have general application. It shall be binding in its entirety and directly applicable in all Member States.' It is important to distinguish between '*4. As the Court has already stated in other contexts, in particular in its judgment of 10 October 1973 (Variola SpA v Amministrazione Italiana delle Finanze[1973] ECR981), the direct application of a Community regulation means that its entry into force and its application in favour of or against those subject to it are independent of any measure of reception intonational law. 5 By virtue of the obligations arising from the Treaty the Member States are under a duty not to obstruct the direct effect inherent in regulations and other rules of Community law. 6 Strict compliance with this obligation is an indispensable condition of simultaneous and uniform application of Community regulations throughout the Community*' (C-50/76—Amsterdam Bulb BV v Produktschap voor siergewassen, para. 4-7); likewise '... *Therefore, by reason of their nature and their function in the system of the sources of Community law, regulations have direct effect and are as such, capable of creating individual rights which national courts must protect*'. See C-43/71—Politi v Ministero delle finanze, ECLI:EU:C:1971:122, para. 9.

[16] Guillem Carrau (2021), p. 353.

which has constitutive effects as said in Chapter 3. This aspect is very well described in AG Opinion in *Feta* case saga[17]:

> 51. As with other industrial property rights, the protection of a name identifying an agricultural product or a foodstuff is conditional upon registration, a compulsory procedure that fulfils objectives similar to those of the Community trade mark, but is unlike the procedure in the wine sector.
>
> 52. It is clear from Article 17(3) that, since it is the sole system of protection for such names in the Community, the requirement of registration must be met even in respect of names which, before the entry into force of the basic registration, were already protected under national laws or were established by usage in Member States with a different system.

Therefore, national systems are still in place to regulate the first step of the application process for EU GI registration and ensure their administrative enforcement (i.e. official controls and verifications of compliance with product specifications established by producers and monitoring of the use of a GI in the market).[18]

This set of tasks implies, among other responsibilities, to examine and publish the GI application at national level to allow any stakeholder to oppose.

4.2.1 Scrutinize the Application and Grant the Opposition Period at National Level

Indeed, the initial design of the registration procedure stated two procedures for the registry, the simplified one and the normal on. Article 17 Regulation (EC) 2081/92 established a simplified procedure for Member States to registry those GIs already protected at national level. After 2003, Regulation (EC) n° 692/2003, the simplified one disappeared as it was foreseen for preexisting GIs at national level and it was not granting an opposition phase at EU level.

When scrutinizing the application, as a matter of fact, Member States must grant that the GI application respects the requirements stated by the legislator and that the Opposition phase is opened at national level. The system does not establish specific procedural safeguards, at EU level, for individuals as said in *PAMPRYL.*[19]

The Court clarified in *La Conqueste*,[20] that the guarantees in favor of individuals were exclusive responsibility of the Member States and there was not margin of maneuver for the European Commission.

> 64. Contrary to what the applicant claims, under the scheme for registering objections established by Regulation 2081/92, the procedural safeguards afforded to individuals fall exclu-

[17] AG Opinion C-465/02—Germany and Denmark v Commission ECLI:EU:C:2005:276, paras. 51–52.

[18] C-35/13, Assica and Krafts Foods Italia, ECLI:EU:C:2014:306, para. 28.

[19] T-114/99—CSR PAMPRYL v Commission, ECLI:EU:T:1999:281, para. 51.

[20] T-215/00—La Conqueste v Commission, ECLI:EU:T:2001:23, para. 64.

sively within the responsibility of the Member States and do not involve the exercise of any discretion by the Commission.

In *La Conqueste*, the applicant was an undertaking established in southwestern France, which produced and hatched mallard duck eggs, and rears and force-feeds the ducks. This stakeholder sent at national level a statement of objections to the GI's registration, *'canard à foie gras du Sud-Ouest'*. The objections were sent as well to the European Commission which pointed out to the applicant that the statement should be sent to the competent French authorities.

In the case at hand, the Court stipulated that the role of the European Commission in the registration proceedings does not exempt the Member State to accomplish with its obligations. The Commission reasoned that this obligation of the Member State implied to publish the GI application at national level to allow any stakeholder to oppose. Any moral or legal person affected by the application had the right to examine the application and, if needed, to oppose. All the oppositions must have been solved at national level before submitting the file to the European Commission as this opposition cannot be done at EU level by the individuals.[21]

In this context, there is not any type of pre-objection procedure for stakeholders from third countries when the file is still at national level. In *Halloumi*,[22] the Court considered that the letters, in which the European Commission explained to the Cyprus Turkish Chamber of Industry that the PDO registration procedure did not contemplate any pre-objection procedure, did not constitute a final decision on the applicant's oppositions.

39. Thus it is clear from the wording of the contested measures that the Commission has not yet taken a final decision on the applicants' oppositions. Furthermore, the contested measures contain no explicit statement that the oppositions are rejected (see, to that effect and by analogy, judgment of 10 July 1990, Automec v Commission, T-64/89, EU:T:1990:42, paragraph 55 and the case-law cited). On the contrary, they explicitly state that no 'formal decision' has been taken by the Commission as regards those oppositions.

40 Thus, only the Commission's final decision adopting, on the basis of Article 52 of Regulation No 1151/2012, implementing acts registering the PDO at issue is capable of bringing about a distinct alteration in the applicant's legal position (see, to that effect and by analogy, order of 10 September 2014, Zentralverband des Deutschen Bäckerhandwerks v Commission, T-354/13, not published, EU:T:2014:775, paragraphs 29 and 30).

41 Illegalities vitiating measures preliminary to the final decision, like the contested measures, which relate to the opposition procedure referred to in Article 51 of Regulation No 1151/2012, are such as to be reasonably raised in support of an action directed against the final measure for which they represent a preparatory step (see, to that effect, judgment of 11 November 1981, IBM v Commission, 60/81, EU:C:1981:264, paragraph 12). In that regard, the applicants' argument that, by the contested measures, they were informed that they would not be invited to the consultations provided for in Article 51(3) of Regulation No 1151/2012 is irrelevant as regards the actionable nature of those measures, since the right of specified third parties to be properly heard, on application by them, during an administrative procedure before the Commission can in principle be given effect to by the Court only

[21] C-447/98 P—Molkerei Großbraunshain and Bene Nahrungsmittel v Commission, ECLI:EU:C:2000:586, C-269/99—Carl Kühne and Others ECLI:EU:C:2001:659.

[22] T-41/16—Cyprus Turkish Chamber of Industry and Others v Commission, ECLI:EU:T:2016:613.

> at the stage of review of the lawfulness of the Commission's final decision (see, to that effect, judgment of 12 October 2011, Association belge des consommateurs Test-Achats v Commission, T-224/10, EU:T:2011:588, paragraph 29, and order of 9 July 2003, Commerzbank v Commission, T-219/01, EU:T:2003:201, paragraph 58; see also judgment of 18 December 1992, Cimenteries CBR and Others v Commission, T-10/92 to T-12/92 and T-15/92, EU:T:1992:123, paragraph 42), and not at the stage of a step in the administrative procedure before the final decision.

The exclusive and exhaustive character of the EU GI system aims at protecting the rights of producers who have made a genuine effort to improve product characteristics, attributes, and quality to be able to use a GI registered under EU law. The national registration procedures are therefore incorporated into the EU decision-making procedure and constitute an essential part thereof. In *Port Charlotte*, the Court said: '*The national registration procedures are therefore incorporated in the decision-making procedure at EU level and constitute an essential part thereof. They cannot exist outside the EU system of protection*'[23]

Currently, the procedure relays on the fact that a producer group can apply for the registration to the authority designated by the Member State. The National stage of the procedure of registration basically covers the following steps: an application is addressed to the competent authority; an examination of the application (product specification, single document[24] and the accompanying documentation) is performed by the competent authority; a national opposition procedure is launched by publishing the application; a submission to the European Commission that happens after the examination of the application for registration and the assessment of the results of any opposition received and any changes to the application agreed with the applicant producer group, and only when the Member State concerned considers that the requirements of the Regulation are met. This national phase must ensure the possibility for any natural or legal person to appeal.

A transitional national protection can be granted at national level[25] that will cease once when the implementing act deciding on the registration enters into force, or the application is withdrawn.[26]

Therefore, the first step of the proceedings is the responsibility for the national authority to design an administrative procedure for the submission and examination of the GIs' applications and to check whether the application fulfils the requirements of the legislator[27]

---

[23] Op. cit. 9, para. 87.

[24] The single document was established in 2006 (Regulation (EC) 510/2006) to grant an homogeneous review of the applications and to offer a greater degree of transparency towards the external stakeholders producing in the same area.

[25] This legal provision is inspired by Joined cases C-129/97 and C-130/97, Chiciak and Fol, ECLI:EU:C:1998:274, paras. 25–26.

[26] See Articles 10 and 11 Regulation (EU) 1141/2024.

[27] Lobato García-Miján (1995), p. 1976.

As regards the applicant, the submission of the candidature for becoming a GI can only be started by a producer group. An applicant producer group shall be an association, irrespective of its legal form, composed of producers of the same product the name of which is proposed for registration. Public bodies and other interested parties may assist in the preparation of the application and in the related procedure, says the legislator. However, a single producer can be considered an applicant producer group if a set of conditions are met. When a GI designates a cross-border geographical area, several applicant producer groups from different Member States or third countries can lodge a joint application for its registration. Such a joint application shall be addressed to all the Member States concerned.[28]

4.2.2 Standard Amendments to the Product Specifications and the Legitimate Interest to Oppose: *Hengstenberg* Case

Amendments to product specifications are a very common issue in GIs. Any evolution of the markets or the technical aspects of elaborating, producing or marketing the protected good can justify that the recognized producer group apply for an amendment. From a procedural point of view, they are classified as Union amendments and standard amendments. The first ones will require an opposition procedure at Union level and the second ones are to be dealt with at Member State or third country level.

Before 2024, the amendments were qualified as minor or major. The procedure relating to an application for a minor amendment to the product specification, set out in the second subparagraph of Article 53(2) of Regulation No 1151/2012 (in force at the time), does not provide for the possibility of lodging an opposition to the proposed amendment, unlike the procedure set out in the provisions regarding an amendment to the product specification which is not minor.

In those circumstances, the action concerning the lawfulness of a decision made by national authorities approving that application for a minor amendment is the only opportunity for natural or legal persons affected by such a decision to oppose it.

In this situation, the producers who do not participate in the scheme have a word in the procedure of modification of the single document, as said in *Spreewälder Gurken* (PGI), also known as *Hengstenberg*.[29] The Court considered that, in so far as the review of the application for registration or for a non-minor amendment of a GI was essentially a matter for the Member State concerned, it was particularly necessary to open widely the possibility for natural or legal persons who might benefit from or, on the contrary, suffer from an economic point of view from the registration or the non-minor amendment of the specification of a product to oppose it so that their arguments could be examined during the national opposition procedure.

[28] See Article 9 Regulation (EU) 1141/2024.

[29] C-53/20, Hengstenberg, ECLI:EU:C:2021:279, para. 39.

The concept of legitimate interest is explained in the context of the procedure applicable to applications for non-minor amendments to the specification of a product covered by a protected geographical indication, as '...*any natural or legal person affected economically, actually or potentially – provided that such an effect is not entirely implausible – by the amendments applied for may establish the 'legitimate interest' required to lodge an opposition to the application for amendment submitted or to bring an action against the decision granting that application, as long as the risk of harm to the interests of such a person is not purely improbable or hypothetical, which is for the referring court to ascertain*'.

This criteria will be verified by Member States according to Article 4 Commission Delegated Regulation (EU) 2025/27.

Currently, any other amendment to a product specification of a GI, that is not a Union amendment, will be considered as a standard amendment. For the purposes of Article 24 of Regulation (EU) 2024/1143, applications for approval of standard amendment to a product specification must be submitted to the authorities of the Member State in whose territory the defined geographical area of the product concerned is located.

Member States may provide that applications for standard amendments are published for opposition at national level. If national opposition is not provided for, and if the application for approval of a standard amendment to a product specification does not come from the applicant producer group that had submitted the application for protection of the name or names to which the product specification refers, the Member State must give that applicant producer group the opportunity to comment on the application, if that applicant producer group still exists.[30]

Any application of a standard amendment must provide a description of the standard amendments and a summary of the reasons for which the amendments were required, and must demonstrate that the proposed amendments qualify as standard.

When requirements are met, the Member State can approve the standard amendment. Where the defined geographical area covers more than one Member State, each Member State concerned shall apply the procedure for the standard amendment separately. The approval decision must include the electronic reference to the published amended consolidated product specification and, where relevant, the amended consolidated single document. When the standard amendment concerns a GI for which a single document has never been published in the Official Journal of the European Union, a single document must always be included.

To conclude this procedure for the standard amendments, the approval decision including the related single document, where relevant, shall be made public by the Member States. The approved standard amendment must be applicable in the Member State concerned following the applicable national rules for entry into force. Likewise, the Member State shall communicate approved standard amendments to the Commission no later than 1 month following the date on which the national approval decision was made public.[31]

[30] Article 4 Commission Delegated Regulation (EU) 2025/27.

[31] Article 5(1) Commission Delegated Regulation (EU) 2025/27.

If an approval of a standard amendment results in an amendment of the single document, the Commission will publish the communication of the standard amendment and the amended single document in the Official Journal of the European Union, C series, within 3 months from the date on which it has received the communication of that standard amendment. When the approved standard amendment does not result in an amendment of the single document, the Commission must make public in the Union register of GIs the communication of the standard amendment within 3 months from the date on which it has received the communication of that standard amendment.[32]

Last but not least, any immediately applicable judicial or administrative decision annulling a decision approving a standard amendment must be communicated to the Commission by the Member State. Moreover, the Member State must transmit to the Commission the consolidated single document and the electronic reference to the product specification, or, in case of an amendment not modifying the single document, only the electronic reference to the product specification, updated following the annulment of that standard amendment. If the standard amendment is annulled by a national decision, the Commission must publish it in the Official Journal of the European Union, C series, or make it public in the Union Register of GI.

4.2.3 Temporary Amendments

Within the standard amendments, there are temporary amendments, as a subcategory that foresees when the modification of the product specification concerns a temporary change resulting from the imposition of obligatory sanitary and phytosanitary measures by the public authorities or a temporary amendment necessary because of a natural disaster or adverse weather conditions, or significant market disturbances due to exceptional circumstances, including geopolitical events, affecting the supply of raw materials, provided that the natural disaster, adverse weather conditions or significant market disturbances are formally recognised by the competent authorities.

According to Article 7 of Commission Delegated Regulation (EU) 2025/27, temporary amendments to a product specification shall be approved and made public by the Member State in whose territory the defined geographical area of the GI concerned is located. Temporary amendments may concern a part of the geographical area. They must be communicated to the Commission together with the reasons supporting them and the national approval decision no later than 1 month following the date on which that decision was made public. A temporary amendment must be applicable in the Member State concerned following the applicable national rules for entry into force.

As said, the main characteristic of the temporary amendments is that they must apply for a limited period set out by the authority approving the amendment. It may only be renewed if the exceptional circumstances referred to in Article 24(5) of Regulation (EU) 2024/1143 since which it was first approved still exist. A renewal

[32] Article 5 (3) and (4) Commission Delegated Regulation (EU) 2025/27.

of temporary amendments must be communicated to the Commission. Where the geographical area covers more than one Member State, each of the Member States concerned must apply the procedure for temporary amendments. Likewise, temporary amendments concerning GIs originating in third countries must be communicated to the Commission, together with the reasons supporting them, by a producer group, either directly or via the authorities of that third country, no later than 1 month following the date on which the temporary amendment was approved.

The communication of an approved temporary amendment to the Commission will be duly made when it contains the elements listed in Article 13 of Implementing Regulation (EU) 2025/26. The Commission must make public the communication of temporary amendments in the Union register of GIs within 3 months from the date on which it has received the communication of temporary amendments.

To conclude, the Member State, third country or producer group that communicated a temporary amendment to the Commission must remain responsible for its content, says the legislator.[33]

4.3 The *Comté* Cheese and the *Piadina*: What Happens When the National Decision Is Reviewed by a National Court

The Commission's decision on registration is subject to the prior decision taken by the competent national authorities on the application for registration[34] but what happens when the National decision is reviewed by a National Court. Any possible annulment of such a decision taken by the national authorities would deprive the Commission's decision of any basis and, consequently, entail a review of the case by the Commission.[35]

Cases like the *Comté*[36] or the *Piadina Romagnola*[37] are helpful to understand that the system has been established within the Common Agriculture Policy and not within IP Law in pure sense and how the division of power among the different Administrations involved in is working on.

As a matter of fact, in *Piadina Romagnola*, GI's standing orders were annulled by a national Court once the opposition phase was opened at EU level.

The General Court had to determine that, with the aim of respecting the roles of the Administrations, the Commission should have considered the annulment by TAR Lazio, in the formal examination of the file and wait for the National court decision become final or not. The General Court found that the Commission could not register the PGI 'Piadina Romagnola' where the Italian authority had approved and forwarded the application to the Commission 6 days after the partial annulment

[33] See Article 7(8) of Commission Delegated Regulation (EU) 2025/27.

[34] C-785/18, GAEC Jeanningros, EU:C:2020:46, para. 36.

[35] See, to that effect, C-454/16 P to C-456/16 P and C-458/16 P, Global Steel Wire and Others v Commission, not published, EU:C:2017:818, para. 31.

[36] Op. cit. 34, para. 38.

[37] T-43/15, CRM v Commission, ECLI:EU:T:2018:208, para. 89.

of the specification by judgment of a lower national court. The Commission approved that PGI and published its decision in the Official Journal of the European Union without waiting for the outcome of the appeal before the Consiglio di Stato (Council of State, Italy) against the judgment of the lower court partially annulling the specification.

On the top of that, this case has been at the center of a dispute concerning its industrial method of production, which the product specification allows alongside artisanal production; this could now also reverberate through the related sales services beyond the traditional kiosks.[38] In fact, it shows an interesting theoretical contrast between two ways of interpreting the function and the essence of EU GIs. Based on this case, it has been identified a "broad approach", more in line with the EU legislator, and a "narrow approach" supported by some social actors, such as the Slow Food Organisation that is aimed at restricting access to GI, and in particular PGI, protection.[39]

As said in the AG Opinion in *Comté*, to avoid the undesirable consequence of an action concerning the lawfulness of a decision made by national authorities approving that application for a minor amendment, it would be appropriate for the Commission to suspend the part of the procedure with which it deals when there are actions pending before national courts against the decision of the national authority which approved the minor amendment of the specification. That would result in proper compliance with the applicants' right to legal protection, on the one hand, and legal certainty, on the other, by preventing the Commission's decisions from being deprived, a posteriori, of a legal basis.[40]

> 69. In summary, the Commission must refrain from dealing with an application for a minor amendment of a PDO specification both if it is aware of the existence of any legal actions pending against the national approval decision and if there are judgments which invalidate that decision in whole or in part. Otherwise, any annulment of that national decision by final judgment of the courts of the Member State, which are still required to give judgment on any actions pending before them, would unexpectedly invalidate the Commission's decision.

In *Comté*, the question raised for a preliminary ruling was whether the EU applicable provisions must be interpreted as meaning that, when the Commission has granted an application made by the authorities of a Member State seeking a minor amendment to the product specification for a PDO, a national court hearing an action concerning the lawfulness of the decision made by those authorities on that application with a view to its submission to the Commission, can decide that there is no longer any need to adjudicate on the dispute before it. The Court stated that, in this case, the national court cannot, on that ground alone, decide that there is no longer needed to adjudicate on the referred dispute pending. To grant an effective judicial protection, there is an obligation to rule on the action. As said by the AG and endorsed by the Court, any possible annulment of such a decision taken by the

[38] Calabrese (2024), p. 350.

[39] Zappalaglio (2015).

[40] Op. cit. 34, AG Opinion, ECLI:EU:C:2019:789, para. 69.

national authorities would deprive the Commission's decision of any basis and, consequently, entail a review of the case by the Commission.[41]

In 2024, the EU legislator established provisions to solve the question of any national challenge to an application for registration. Article 16 Regulation (EU) 2024/1143 provides that the Member States shall inform the Commission of any national administrative or judicial proceedings that may prejudice the GIs registration. As regards applications for protection and judicial challenges submitted in the national procedure, the Commission should be informed in a timely and regular manner where procedures are launched before national courts or other bodies concerning an application for registration forwarded by the Member State to the Commission, and of the results of those procedures. For the same reason, where a Member State considers that a national decision on which the application for protection is based is likely to be invalidated because of national judicial proceedings, it should inform the Commission of that assessment.

Then, if the Member State requests the suspension of the examination of an application at Union level, the Commission should be exempted from the obligation to meet the deadline for examination. In order to protect the applicant from vexatious legal actions and to preserve the applicant's right to secure the protection of a name within a reasonable time, the exemption should be limited to cases in which the application for registration has been invalidated at national level by an immediately applicable but not final judicial decision or in which the Member State considers that the action to challenge the validity of the application is based on valid grounds.

The exemption provided must have effect until the Commission is informed by the Member State concerned that the original application has been restored or that the Member State withdraws its request for suspension. If the decision of applying has been invalidated in full or in part by a final decision taken by a national court, the Member State shall consider appropriate action such as withdrawal or modification of the application for registration at Union stage, as necessary.

Consequently, these set of provisions will avoid situations like the one described in *Piadina Romagnola* and *Comté*.

4.4 The Role of the European Commission: *Kühne* and *Jambon sec de Corse*

4.4.1 Verification of the Application

The landmark case regarding the role of the European Commission in the GIs registration was the *Kühne* case, which is also known as the pickled gherkins case (*Spreewälder Gurken*).[42] Its importance relays on the fact that it can be extracted

[41] Op. cit. 34, paras. 39–40.

[42] Op. cit. 10, paras. 53–54.

from the case that the verification task of the Commission cannot consist of a mere formal control.

In this preliminary ruling, there is a clear answer to the question about the competences of the European Commission (it is related to the, at that time, existing 'simplified' procedure):

> 53. It follows that the decision to register a designation as a PDO or as a PGI may only be taken by the Commission if the Member State concerned has submitted to it an application for that purpose and that such an application may only be made if the Member State has checked that it is justified. That system of division of powers is attributable particularly to the fact that registration assumes that it has been verified that a certain number of conditions have been met, which requires, to a great extent, detailed knowledge of matters particular to the Member State concerned, matters which the competent authorities of that State are best placed to check.

The Court considered that, under that system of division of powers, it is for the Commission, before registering a designation in the category applied for, to verify, in particular, first, that the specification which accompanies the application complies with the applicable Regulation, that is to say that it contains the required information and that that information does not appear to contain obvious mistakes, and, secondly, on the basis of the information contained in the specification, that the designation satisfies the requirements of the applicable regulation, at that time Article 2(2)(a) or (b) of Regulation No 2081/92. To sum up, according to the Court, the Commission might base its assessment on that previously made by the national authorities where that assessment is not vitiated by a manifest error.[43]

AG conclusions in *Comté*[44] can be quoted again because they contain a clear explanation of a problem which solution was not clear: the annulment of the national act which approves the minor amendment and is then forwarded to the Commission would lead to a *chain reaction* in the form of depriving the Commission's decision of a legal basis. AG considers that to avoid that undesirable consequence, it would be appropriate for the Commission to suspend the part of the procedure with which it deals when there are actions pending before national courts against the decision of the national authority which approved the minor amendment of the specification. That would result in proper compliance with the applicants' right to legal protection, on the one hand, and legal certainty, on the other, by preventing the Commission's decisions from being deprived, a posteriori, of a legal basis.[45]

---

[43] Op. cit. 10, para. 60.

[44] Op. cit. 34.

[45] The judgment of the General Court of 23 April 2018, CRM v Commission (T-43/15, not published, EU:T:2018:208), found that the Commission could not register the PGI 'Piadina Romagnola' where the Italian authority had approved and forwarded the application to the Commission 6 days after the partial annulment of the specification by judgment of a lower national court. The Commission approved that PGI and published its decision in the Official Journal of the European Union without waiting for the outcome of the appeal before the Consiglio di Stato (Council of State, Italy) against the judgment of the lower court partially annulling the specification.

In particular, the AG justifies this reasoning in the obligation to ensure the effective legal protection of persons who have brought actions before the national courts, whose right of action is based on a provision of EU law, and, possibly, the principle of sound administration,[46] should lead the Commission to include in its assessment of an application any actions which have been brought against the national decision before the courts of the Member State concerned.

In *Kühne,* the Court does not say that the Commission was limited by the assessment of the national authorities on, among other aspects, the compatibility of the name at issue with EU law.

This question has been consistently analyzed by the Court in *Jambon sec de Corse* case.[47] As said, the Court was again set to decide on the scope of European Commission's competences when examining GI applications. In the case at hand, the Consortium des Charcutiers Corses seek to have set aside the judgment of the General Court[48] by which the General Court dismissed their action seeking the annulment of Commission Implementing Decision (EU) 2021/1879 of 26 October 2021 rejecting three applications for protection of a GI ['Jambon sec de l'Île de Beauté' (PGI), 'Lonzo de l'Île de Beauté' (PGI), 'Coppa de l'Île de Beauté' (PGI)].

The names "Jambon sec de Corse"/"Jambon sec de Corse – Prisuttu", "Lonzo de Corse"/"Lonzo de Corse – Lonzu" and "Coppa de Corse"/"Coppa de Corse – Coppa di Corsica" were registered as PDOs by the European Commission in 2014.[49] In 2018, France approved the following names, used by the appellants in the case at hand: "Jambon sec de l'Île de Beauté", "Coppa de l'Île de Beauté", "Lonzo de l'Île de Beauté", "Saucisson sec de l'Île de Beauté", "Pancetta de l'Île de Beauté", "Figatelli de l'Île de Beauté" and "Bulagna de l'Île de Beauté" with a view to forwarding them to the European Commission for their approval. After several exchange of communications, the Commission rejected the applications for registration of the names "Jambon sec de l'Île de Beauté", "Lonzo de l'Île de Beauté" and "Coppa de l'Île de Beauté", on the ground that, since those names had been used in breach of Article 13 of Regulation No 1151/2012, those applications did not comply with the eligibility criteria for registration, namely Article 7(1)(a) of that regulation. This decision was appealed by the referred Consortium and, as said before, the action was dismissed in its entirety.

The General Court recalled, in paragraph 38 of that judgment, that it is for the Commission, under Article 52(1) of that regulation, to refuse registration if it considers the use, in trade, of the name—the registration of which is sought—to be unlawful. Next, in paragraphs 42 to 44 of the judgment, the General Court stated

---

[46] Article 41 of the Charter, to which the General Court referred in its judgment of 23 April 2018, *CRM* v *Commission* (T-43/15, not published, EU:T:2018:208, paras. 67–83).

[47] C-579/23 P, Consortium des Charcutiers Corses ECLI:EU:C:2024:832.

[48] T-34/22, Cunsorziu di i Salamaghji Corsi—Consortium des Charcutiers Corses and Others v Commission EU:T:2023:386.

[49] Commission Implementing Regulation (EU) No 581/2014 (OJ 2014 L 160, p. 23), Commission Implementing Regulation (EU) No 580/2014 (OJ 2014 L 160, p. 21) and Commission Implementing Regulation (EU) No 582/2014 (OJ 2014 L 160, p. 25).

that the Commission had a margin of independent discretion when scrutinising applications for registration. The General Court found, in paragraph 49 of that judgment, that the Commission had indeed scrutinised the applications for registration.

At the stage of the Court of Justice, the appeal was dismissed since for the Court the General Court correctly concluded that the Commission has independent discretion, so a decision of a national court which has the force of res judicata cannot be relied upon in order to call into question the Commission's assessment; and, on the top of that,the Court of Justice considered that the General Court did not err in law in finding that there was a risk of evocation between a name protected as a PDO and the names for which registration as a PGI was sought. In the view of the Court, according to Article 50(1) of Regulation No 1151/2012, for the Commission was to check whether applications for registration were justified and met the conditions laid down in that regulation and by stating that it was in no way apparent from that provision that the scrutiny of the conditions for registration by the Commission was limited, in any way, by the initial scrutiny carried out by the national authorities. Consequently, the Court endorsed that the Commission was not bound by the assessment of the national authorities and that it had a margin of independent discretion as regards its decision to register a name as a PGI[50]:

> 39 In the second place, that independent discretion on the part of the Commission is supported, first, by a reading of recital 19 of Regulation No 1151/2012, which states that ensuring uniform respect throughout the European Union for intellectual property rights related to names protected in the European Union is a priority that can be achieved more effectively at EU level. As the Advocate General observes in point 63 of his Opinion, the Commission is vested, in the context of the attainment of that objective, with the power to ensure the uniform application of the conditions for registration, so that, in the absence of independent discretion in that regard, the Commission would not be able to prevent those conditions for registration from being applied differently within the Member States.

It follows that the Commission must grant the absence of manifest errors and that the situation of the stakeholders outside of the Member State have been taken into consideration:

> 40 Second, recital 58 of Regulation No 1151/2012 states that the applications should be examined by the national authorities, in compliance with minimum common provisions, including a national objection procedure. According to that recital, the Commission should subsequently scrutinise applications to ensure that there are no manifest errors and that EU law and the interests of stakeholders outside the Member State of application have been taken into account.

Likewise, the Court interpreted that there is a clear mandate for the Commission to 'scrutinise' and verify the applications:

[50] As is apparent from paragraph 59 of that judgment, the Commission has 'limited, if any' discretion only as regards the assessment of the legality of the documents constituting the file relating to the application for registration gathered by the national authorities (see, to that effect, op. cit. 25, paras. 25, 26 and 36).

> 41 If the scrutiny of applications for registration by the Commission were to be limited, as the appellants claim, to verifying that the files are complete and that there are no manifest errors, that institution would be unable to 'scrutinise' the applications as provided for by recital 58 of Regulation No 1151/2012. In the context of scrutinising the applications, the Commission cannot be prevented from verifying whether that application complies with the conditions for registration laid down by that regulation, as required by Article 50(1) of that regulation. That is borne out by the significance which the wording of that recital attaches to compliance both by the application for registration and by the national authorities with the minimum common requirements resulting in particular from that regulation.

Therefore, the final responsibility of the registration falls within the action of the Commission. A decision which will be independent and not influenced by the review of the application done by the national administration:

> 42 In addition, it should be noted that, according to recital 61 of that regulation, the Commission should be responsible for decision-making on registration. The Commission can assume final responsibility for decision-making on registration only if it has independent discretion, without being bound by the outcome of the preliminary scrutiny of the conditions for registration carried out by the national authorities.

Another interesting outcome from *Jambon sec de Corse* is the extent of the powers of the Commission in assessing the applications by the way of identifying that it is for the European Commission to assess the nature and the scope of and develop the 'appropriate means' without being bounded by the scrutiny carried out by the national authorities.

For instance, as regards the risk of evocation, the Judgment contains an example of which is the role of the European Commission which consist of taking into account of all the relevant factors surrounding the use of the name in question:

> 80. It should be borne in mind that the risk of evocation is established where the use of a name creates a sufficiently clear and direct link between that name and the PDO in the mind of an average European consumer who is reasonably well informed and reasonably observant and circumspect. In that regard, the existence of such a link may arise from several factors, namely, in particular, the partial incorporation of the protected designation, the phonetic and visual resemblance of the two names and the resulting similarity, and, even in the absence of those factors, the conceptual proximity between the PDO and the name at issue or the similarity between the products covered by that PDO and the products or services covered by that name. In the context of the assessment carried out by the Commission, it must take account of all the relevant factors surrounding the use of the name in question (judgment of 9 September 2021, Comité Interprofessionnel du Vin de Champagne, C-783/19, EU:C:2021:713, paragraph 66).

Finally, the appeal against the judgment was filed by the *Syndicat des charcuteries* was dismissed by the Court.[51]

Nowadays, the legal framework reflects an expanded set of competences for the European Commission as regards the registration procedure. The Commission,

[51] T-597/23, Syndicat de défense and de promotion des charcuteries corses v Commission, ECLI:EU:T:2025:844.

following Article 15 Regulation (EU) 2024/1143, shall examine the application for registration; in particular, that it contains the required information and that they do not contain manifest errors, considering the outcome of the national examination and opposition procedure carried out by the Member State concerned. In a nutshell, the application for registration must contain:

(a) the single document;
(b) the accompanying documentation referred to, where relevant, information explaining any proposed limitations on the use or on the protection of the geographical indication, and any transitional measures, proposed by the applicant producer group, the name and contact details of the applicant producer group and the name and contact details of one or more of the competent authorities, delegated or product certification bodies or natural persons verifying compliance;
(c) a declaration by the Member State to which the application was addressed at the national stage of the procedure of registration, confirming that the application meets the conditions for registration;
(d) any transitional period granted or proposed by the national authorities following the national examination and opposition procedure as well as information on the related admissible oppositions; and
(e) the electronic publication reference to the up-to-date product specification.

The Commission must check, in accordance with Article 15 of Regulation (EU) 2024/1143, if an application for registration is complete and if it has been submitted in accordance with Article 10(6), Article 13(4) and Article 14 of Regulation (EU) 2024/1143 and with Article 35 of Commission Implementing Regulation (EU) 2025/26.

The application will be considered complete if it includes all the elements required for an application in accordance with Article 13(1) of Regulation (EU) 2024/1143 and complies with Article 2(1) of Commission Implementing Regulation (EU) 2025/26.[52]

For instance, the single document will be considered complete if it includes all the information required in Article 50(1) of Regulation (EU) 2024/1143, or Article 95 of Regulation (EU) No 1308/2013 or Article 23 of Regulation (EU) 2019/787 for agricultural products, wine and spirit drinks, respectively.[53] As regards the geo-

[52] Article 3(2) Commission Implementing Regulation (EU) 2025/26.

[53] According to Article 4 Commission Implementing Regulation (EU) 2025/26, the single document of GIs for agricultural products and spirit drinks shall be concise and not exceed 2500 words. For geographical indications for wine it shall not exceed 5000 words. Those thresholds may be exceeded in duly justified cases. The single document shall indicate the classification of the product in accordance with the Combined Nomenclature heading and code, as referred to in Article 6(1) of Regulation (EU) 2024/1143 for all sectors. Where the single document includes specific requirements on packaging and labelling or other applicable requirements, provided for by the product specification, any restriction resulting from those requirements shall be accompanied by a summary of the product-specific justification contained in the correspondent product specification.

graphical area, it must be defined in a precise way that presents no ambiguities, referring as far as possible to physical or administrative boundaries. Maps may be added to the application.[54]

When the application for registration of a name or approval of an amendment describes two or more distinct products which are entitled to use that name, compliance with the requirements for registration will be shown separately for each such product. For the legislator, 'distinct products' mean products that, although using the same registered name, are differentiated when placed on the market or considered as different products by consumers.[55]

If all applicable provisions are fulfilled, the Commission will publish in the Official Journal of the European Union the single document and the reference to the publication of the product specification.[56]

Along the procedure, changes can be made to the product specification. The Member State shall update the single document and, if they are substantial, thus affecting interests that had not been considered in the national opposition procedure carried out in accordance with Article 10(4) of Regulation (EU) 2024/1143, those changes shall be subject to an additional national opposition procedure.

The publication of the single document in the Official Journal opens the 'Union opposition procedure' that contains a second set of tasks for the European Commission as dealing with the oppositions and notice of comments and, if proceeds, to the publication and registration of the GI in cause.

4.4.2 Opposition Period

According to Article 17 Regulation (EU) 2024/1143, any natural or legal person having a legitimate interest and established or resident in a Member State other than the one from which the application for registration at Union stage was submitted may lodge an opposition with the Member State, in which it is established or resident.

Member States, third countries authorities or producer groups established or resident in a third country shall ensure that the single document is a faithful summary of the product specification and that there is no substantial divergence between them. Where an inconsistency is found after the registration of the geographical indication, the Member State, third country or the producer group established or resident in a third country that submitted the application shall take the necessary measures to remedy that inconsistency. Names of natural or legal persons that are included in the single document shall be published.

[54] Article 6 Commission Implementing Regulation (EU) 2025/26.

[55] It may also refer to agricultural products covered by different classifications under the Combined Nomenclature referred to in Council Regulation (EEC) No 2658/87 (Council Regulation (EEC) No 2658/87 of 23 July 1987) on the tariff and statistical nomenclature and on the Common Customs Tariff (OJ L 256, 7.9.1987, p. 1) or to wines and spirit drinks covered by different categories listed in Part II of Annex VII to Regulation (EU) No 1308/2013 or Annex I to Regulation (EU) 2019/787, respectively.

[56] See Articles 9, 10, 12, 13, 28, 29, 30, 31, 46 and 47, Article 48(1) and (2) and Article 50 of Regulation (EU) 2024/1143.

Among other requirements, an opposition must contain a statement formally signifying the opposition to the registration of that name.[57] The European Commission must inform the applicant about the opposition without delay.[58]

When the Commission receives an admissible opposition, the acts to be adopted will follow Article 21 (3) Regulation (EU) 2024/1143 with the involvement Quality Policy Committee for agricultural products, wine and spirit drinks.

Once concluded the opposition is admissible,[59] the Commission will invite the opponent and the applicant to engage in appropriate consultations for a reasonable period that shall not exceed 3 months as referred to in Article 17(4) of Regulation (EU) 2024/1143.

Simultaneously, within 3 months from the date of publication of the single document and the reference to the publication of the product specification, the authorities of a Member State or of a third country, or a natural or legal person established or resident in a third country can submit to the Commission a notice of comments. In terms of Article 18 Regulation (EU) 2024/1143, the aim of the notice of comments is to point out any error or contain additional information in relation to the application for registration, including possible infringement of Union law. A notice of comments shall not confer any rights on the authorities or persons, nor trigger an opposition procedure.

In the absence of an admissible opposition, the Commission shall, by means of implementing acts, register the GI.

Lastly, an opposition period will be opened in case the Member States consider that the consequent changes to the application for registration referred to in Article 17(6) of Regulation (EU) 2024/1143 were substantial, thus affecting interests that had not been considered in the national opposition procedure carried out in accordance with Article 10(4) of that Regulation.[60]

---

[57] Article 9 Commission Implementing Regulation (EU) 2025/26.

[58] Article 2(1) Commission Delegated Regulation (EU) 2025/27.

[59] See Article 19 Regulation (EU) 2024/1143 for the admissibility criteria '*An opposition lodged in accordance with Article 17 shall be admissible only if the opponent shows that:* (a) *the proposed geographical indication does not comply with the definition of the geographical indication or with the requirements referred to in this Regulation, Section 2 of Chapter I of Title II of Part II of Regulation (EU) No 1308/2013 or Article 3(4) and Chapter III of Regulation (EU) 2019/787 as applicable;* (b) *registration of the proposed geographical indication would be prevented by one or more of the circumstances referred to in Article 28, Article 29, Article 30 or Article 48(1); or* (c) *the registration of the proposed geographical indication would jeopardise the existence of a wholly or partially identical name or of a trade mark or the existence of products which have been legally on the market for at least five years preceding the date of the publication of the information provided for in Article 15(4). 2. The admissibility of an opposition shall be assessed by the Commission in relation to the territory of the Union*'.

[60] Article 2(4) Commission Delegated Regulation (EU) 2025/27.

4.4.3 Union Amendments to a Product Specification

The Commission is also responsible for the amendments to a product specification that are qualified 'Union amendments', requiring an opposition procedure at Union level[61] and likewise for the cancellation of the registration.[62] According to Article 24 (3) Regulation (EU) 2024/1143, an Union amendment entails a change of the single document or its equivalent and:

(a) includes a change:
   a. for agricultural products, in the name or in the use of the name;
   b. for wine, in the name or in the use of the name, or in the category of product or products designated by the geographical indication;
   c. for spirit drinks, in the name or any part of the name or in the use of the name, or in the category of product or products designated by the geographical indication, or in the legal name; or
(b) risks voiding the link to the geographical area referred to in the single document; or
(c) entails further restrictions on the marketing of the product.

The Commission checks, in accordance with Article 15 of Regulation (EU) 2024/1143, if an application for approval of a Union amendment to a product specification is complete and if it has been submitted in accordance with Article 10(6), Article 13(4) and Article 14 of Regulation (EU) 2024/1143, and with Article 35 of Commission Implementing Regulation (EU) 2025/26. An application for approval of a Union amendment to a product specification will be considered complete if it includes all the elements referred to in Article 10(1) to (4), where appropriate, and complies with Article 10(7) of Commission Implementing Regulation (EU) 2025/26.[63]

As explained before, although standard amendments must be assessed and approved by Member States or third countries in whose territory the geographical area of the product concerned is located and communicated to the Commission, in the Commission hands relays the responsibility of making those amendments public.[64]

An application for approval of a Union amendment must include a single document if the single document or equivalent has never been published in the Official

---

[61] Article 24(3) Regulation (EU) 2024/1143 states that an amendment shall be considered as a Union amendment if it entails a change of the single document or its equivalent and: (a) includes a change: (i) for agricultural products, in the name or in the use of the name; (ii) for wine, in the name or in the use of the name, or in the category of product or products designated by the geographical indication; (iii) for spirit drinks, in the name or any part of the name or in the use of the name, or in the category of product or products designated by the geographical indication, or in the legal name; or (b) risks voiding the link to the geographical area referred to in the single document; or (c) entails further restrictions on the marketing of the product. The criteria referred to in points (a), (b) and (c) shall be verified by Member States.

[62] Article 25 of Regulation (EU) 2024/1143.

[63] Article 11 Commission Implementing Regulation (EU) 2025/26.

[64] Article 12 y 13 Commission Implementing Regulation (EU) 2025/26.

Journal of the European Union. The single document must comply with Article 50(1) of Regulation (EU) 2024/1143, Article 95 of Regulation (EU) No 1308/2013 or Article 23 of Regulation (EU) 2019/787, depending on whether the GI designates an agricultural product, a wine or a spirit drink, respectively.[65]

4.4.4 Cancellation of the Registration

Furthermore, to conclude its list of tasks, the Commission holds the responsibility of cancel the registration of a GI in certain circumstances. According to Article 25 Regulation (EU) 2024/1143, the Commission can, on its own initiative or on a duly substantiated request by a Member State, a third country or any natural or legal person having a legitimate interest and established or resident in a third country, by means of implementing acts, cancel the registration of a GI in the following cases:

(a) where compliance with the requirements for the product specification can no longer be ensured; or
(b) where no product has been placed on the market under the geographical indication for at least the preceding seven consecutive years.

The opposition procedure is applied *mutatis mutandis* to the cancellation of the registration. As regards the admissibility of an opposition to the cancellation, the opponent to the cancellation must show continued commercial reliance on the registered name by an interested natural or legal person by a statement explaining the legitimate interest[66]

The Commission must examine the request to cancel the registration of the GI. In particular, the Commission must check, in accordance with Article 15 of Regulation (EU) 2024/1143, if a request of cancellation of the registration of a GI is complete and if it has been submitted in accordance with Article 10(6), Article 13(4) and Article 14 of Regulation (EU) 2024/1143 and with Article 35 of Commission Implementing Regulation (EU) 2025/26.

A request for cancellation of the registration of a GI will be considered complete if it includes all the elements referred to in Article 14(1) or (6) and complies with Article 14(2) or (7) of Commission Implementing Regulation (EU) 2025/26.[67] If the request does not fall into either of the two cases referred, it will be rejected. The measure provided for in this decision, under Regulation (EU) No 1151/2012, requested accordance with the opinion of the Committee for agricultural product quality policy.[68]

---

[65] Article 3 Commission Delegated Regulation (EU) 2025/27.

[66] Article 14 Commission Implementing Regulation (EU) 2025/26.

[67] Article 15 Commission Implementing Regulation (EU) 2025/26.

[68] See for instance, Commission Implementing Decision of 14 November 2013 concerning the rejection of a request to cancel a name entered in the register of protected designations of origin and protected geographical indications provided for in Regulation (EU) No 1151/2012 of the

Where the registration of a GI has been cancelled, the Union register of GIs will show the name as cancelled from the date on which the relevant implementing act takes effect. That register will maintain a record of the cancellation, including the electronic reference to the decision of cancellation.[69]

Once the cancellation has been completed, the legislator impedes an immediate registration of the term as trade marks. The registration of the name as an intellectual property right other than a GI, in particular as a trade mark, will be prohibited for one year after the cancellation of the GI, unless such intellectual property right had existed, or such a trade mark had been registered, before the registration of the GI.[70]

4.4.5 Rejection of the Application

The rejection of an application also forms part of the tasks performed by the Commission. It consists of a Commission implementing decision that explains the justification of the refusal in compliance with the opinion of the Committee for agricultural product quality policy. Among others, applications as *Choucroute d'Alsace* (PGI), *Džiugas* (PGI); *Germantas* (PGI); *Eilenburger Sachsenquelle* (PDO), *Eilenburger Sanusquelle* (PDO) and *Pomazánkové máslo* (TSG) were rejected by the European Commission.[71]

For instance, in 1994, Germany notified to the Commission many applications for registration concerning mineral waters pursuant to Article 17(2) of Council Regulation (EEC) No 2081/92, 108 of which were still pending in 2010. In the case of the 31 names listed in Annex I to the Decision, the name proposed for registration

European Parliament and of the Council (Kołocz śląski/kołacz śląski (PGI)) (notified under document C(2013) 7626).

[69] Article 16(7) Commission Implementing Regulation (EU) 2025/26.

[70] See Article 22 Regulation (EU) 2024/1143.

[71] See among other: Commission Decision of 12 January 2006 concerning an application for registration in the Register of protected designations of origin and protected geographical indications provided for in Council Regulation (EEC) No 2081/92 (Choucroute d'Alsace) (PGI) (notified under document number C(2006) 5). [See—OJ C 206, 2.9.2003, p. 2]; Decision of 19 December 2008 rejecting an application for entry in the register of protected designations of origin and protected geographical indications provided for in Council Regulation (EC) No 510/2006 (Džiugas) (PGI) (notified under document number C(2008) 8423); Commission Decision of 19 December 2008 rejecting an application for entry in the register of protected designations of origin and protected geographical indications provided for in Council Regulation (EC) No 510/2006 (Germantas) (PGI) (notified under document number C(2008) 8430); Commission Implementing Decision of 25 July 2011 rejecting two applications for entry in the register of protected designations of origin and protected geographical indications provided for in Council Regulation (EC) No 510/2006 (Eilenburger Sachsenquelle (PDO)), (Eilenburger Sanusquelle (PDO)) (notified under document C(2011) 5251). [See OJ C 54, 7.3.2009, p. 7]; Commission Implementing Decision of 13 November 2013 rejecting an application for entry in the register of traditional specialities guaranteed provided for in Regulation (EU) No 1151/2012 of the European Parliament and of the Council (Pomazánkové máslo (TSG)) (notified under document C(2013) 7615).

was not included in the list of natural mineral waters recognized by Member States in accordance with Article 1 of Directive 2009/54/EC of the European Parliament and of the Council of 18 June 2009 on the exploitation and marketing of natural mineral waters. Therefore, these names could not be recognized in the internal market as marketable mineral waters and should accordingly not be registered. In the case of the 7 names listed in Annex II to the Decision, only the name and no further information had been submitted, thus preventing the Commission to perform the scrutiny required pursuant to Article 6 of Regulation (EC) No 510/2006 to assess whether these names meet the conditions for registration. Therefore, the Commission decided that these names should not be registered.

In the case of the 70 names included in the Annex III to the Decision, the Commission asked the German authorities, by a letter on 20 July 2004, to provide additional information, on details of control bodies, willingness of producers to cover inspection costs and structures guaranteeing the marketing of the water under a single name. On 15 May 2006 and again on 22 May 2007 the European Commission asked the German authorities to either reply to the letter of 20 July 2004 or to withdraw the pending applications and informed them that otherwise the European Commission would consider a rejection of these applications. As no additional information was provided by Germany, the Commission concluded that these names should not be registered. In the light of the above, the applications for registration of the designations listed in the Annexes to the Decision were rejected.[72]

4.4.6 Establishing and Maintaining a Union Registry of Geographical Indications

The Commission oversees establishing and maintaining a publicly accessible Union register of GIs with three parts corresponding to GIs of wine, of spirit drinks and of agricultural products respectively. The EUIPO must maintain and keep up to date the Union register with respect to registrations, amendments, and cancellations of GIs.[73]

The right to appeal an implementing regulation that registers, amends or cancels a GI has been subject of discussion in the past because natural or legal persons may claim that a contested provision is of individual concern to them only if it affects them by reason of certain attributes which are peculiar to them or by reason of circumstances in which they are differentiated from all other persons.[74] GIs concerning products from third countries that are protected in the Union under an international

[72] Commission Decision of 19 April 2010 rejecting a list of applications for entry in the register of protected designations of origin and protected geographical indications provided for in Council Regulation (EC) No 510/2006 (notified under document C(2010) 2385), OJ C 54, 7.3.2009, p. 7.

[73] See Article 22 Regulation (EU) 2024/1143.

[74] See, for instance, C-309/89—Codorniu v Council ECLI:EU:C:1994:197 paras. 20–21; and T-100/94—Michailidis and Others v Commission, ECLI:EU:T:1998:197, paras. 67–69.

agreement to which the Union is a contracting party may be also entered in the Union register.

The Union register of GIs referred to in Article 22 of Regulation (EU) 2024/1143 established as a digital solution allowing the technical storage of and public access to all entries concerning GIs, including applications for registration, for Union amendment and for cancellation, rejections, publications for opposition, registrations, approvals of Union amendments, publications of standard and temporary amendments, and cancellations.

As it will be described in the following subsection, the Commission is the owner of that Union register. The digital solution shall be hosted and managed by the EUIPO, in accordance with Article 22(2) of Regulation (EU) 2024/1143. The Commission will make available the relevant data to EUIPO. The Union register must be available in all the official languages of the Union.[75]

As to the controls and enforcement, that will be analyse afterwards, there is an additional role for the Commission which consists of auditing the activity of the Member States based on a risk analysis.

4.5 Duties and Competence of the EUIPO in Cooperation with DG AGRI

The competences of EUIPO in the field of GIs procedures have their roots in a Memorandum of Understanding signed on 25 April 2019 between the European Commission and the EUIPO which provided a framework for the conclusion of detailed working arrangements with individual Commission departments, and specifically envisaged in its Chapter 7 deepened practical cooperation between EUIPO and DG AGRI. Under that framework, on 17 December 2019, EUIPO and DG AGRI signed an Administrative Agreement and have been cooperating closely on matters relating to GIs, such as pre-assessment of GI applications for agricultural products, wines and spirit drinks, and have launched GI View in 2020 to facilitate access to information concerning GIs.

By virtue of Regulation (EU) 2024/1143, as advanced before, the EUIPO is entrusted with new tasks in the management of GIs for agricultural products, wines and spirit drinks, namely maintaining and keeping up to date the Union register with respect to registrations, amendments and cancellations and establishing and managing a domain name information and alert system.

A revised Administrative Agreement entered into force on 1st January 2025 and replaced the Administrative Agreement signed on 17 December 2019, and subsequent amendment in its entirety.[76]

The revised agreement reinforces key objectives of the collaboration between both entities, which includes cooperation in the field of GIs and the GI View

[75] Article 16 Commission Implementing Regulation (EU) 2025/26.

[76] The 2024 Agreement is accessible following this link https://agriculture.ec.europa.eu/system/files/2022-04/administrative-agreement-between-dgagri-euipo-171219_en_0.pdf.

database,[77] as well as collaborating on communication, training activities and enforcement matters.

Under Chapter 2 of the revised agreement, the EUIPO is responsible for maintaining and keeping up to date the Union register GIs for agricultural products, wines and spirit drinks. Without prejudice to the rights and obligations set out in Regulations (EU) 2023/2411 and (EU) 2024/1143, as well as in the subsequent implementing and delegated regulations, EUIPO and DG AGRI established channels of communication to exchange views, on a regular basis or if need arises, with a view to discussing technical aspects and addressing issues relating to the setup and maintenance of a single Union register for GIs.

Both entities have agreed on stand ready to cooperate on the development of a domain name information and alert system, within the framework to be established by delegated acts giving effect to Article 35(2) of Regulation (EU) 2024/1143, notably by exploring the option to extend the existing domain name alert and information system with the EURid related to trade marks to GIs as well as other European top level domain name registries.

As regards GI View, Article 3 of the Agreement defined that it is the comprehensive electronic database with advanced search capabilities containing registered and extended data, beyond those included in the Union register, displayed in a user-friendly format, in relation to:

- GIs and GI applications for agricultural products, wines and spirit drinks, subject to the provisions of the EU Regulations;
- GIs pertaining to products of non-EU countries, protected in the EU under bilateral, regional and multilateral international agreements entered into by the Union;
- GIs, pertaining to products of EU Member States, protected in non-EU countries under bilateral, regional and multilateral international agreements entered into by the Union.

GI View also will provide a link with EUIPO's Intellectual Property Enforcement Portal ('IPEP') with a view to facilitating enforcement actions for GI stakeholders. The revised Agreement confirmed that EUIPO, in collaboration with DG AGRI, will continue to provide technical assistance and training to the competent authorities of the Member States and third countries on how to utilise the tool, as required.

A dedicated chapter, number 3, outlines the cooperation between EUIPO and DG AGRI in examination practices for GIs, pre-assessment of applications, cooperation with WIPO and the Geneva Act, and the establishment of GI guidelines both for agricultural as well as for craft and industrial GIs.

Without prejudice to Regulation (EU) 2024/1143, the cooperation continues in the pre-assessment of GI applications for the agricultural sector. Pre-assessment includes GI applications for agricultural products, wines and spirit drinks under the respective EU Regulations. It includes also GI files received under bilateral, regional and multilateral agreements, including the Geneva Act, as deemed

[77] See https://www.tmdn.org/giview/.

necessary by DG AGRI. Pre-assessment will be limited to EUIPO providing a non-binding opinion on GI applications it receives from DG AGRI. The EUIPO will deliver the opinion within the agreed timeline and in the format requested by DG AGRI. In any case, the revised Agreement declares that DG AGRI remains fully responsible for the decision taking with respect to GI applications for the agricultural sector as this remains the exclusive competence of the Commission.

Finally, it is for DG AGRI to inform the European Parliament and the Council on a yearly basis on the assistance it has received from the EUIPO in the exercise of its tasks over the course of the previous calendar year.

4.6 Controls and Enforcement

Despite many efforts to develop GIs strategies and roadmaps, infringements in the markets are still rampant. Enforcement and controls are key for the effective implementation of GIs at each stage of the value chain (e.g. production, preparation, distribution, placing on the market): from the compliance with the product specification (the upstream market) to the respect of GIs as Intellectual Property Rights placed on the market and the provision of trustworthy information to consumers (downstream market).[78]

The framework for GIs controls and enforcement as defined in the EU law, is a combination of specific rules: the Official Control Regulation (OCR) (EU) No 2017/625[79] and the Directive on Enforcement of Intellectual Property Rights (IPRED) 2004/48/EC.[80] The OCR applies to official compliance controls on the use and labelling of PDOs, PGIs and TSGs while IPRED requires Member States to provide for measures, procedures and remedies to ensure that IPR are enforced on the downstream market.

The most frequent organisation in Member States is a central administration in charge of supervising the whole control procedure over the value chain, with some exceptions of regionalised organisations (for instance in Spain). The central competent authority (CCA) may delegate parts of responsibility to other competent

[78] Commission Staff working document Evaluation of GIs and TSG in the EU, SWD/2021/0427 final, accessible in https://eur-lex.europa.eu/legal-content/EN/TXT/?uri=SWD:2021:427:FIN.

[79] Regulation (EU) 2017/625 of the European Parliament and of the Council of 15 March 2017 on official controls and other official activities performed to ensure the application of food and feed law, rules on animal health and welfare, plant health and plant protection products, amending Regulations (EC) No 999/2001, (EC) No 396/2005, (EC) No 1069/2009, (EC) No 1107/2009, (EU) No 1151/2012, (EU) No 652/2014, (EU) 2016/429 and (EU) 2016/2031 of the European Parliament and of the Council, Council Regulations (EC) No 1/2005 and (EC) No 1099/2009 and Council Directives 98/58/EC, 1999/74/EC, 2007/43/EC, 2008/119/EC and 2008/120/EC, and repealing Regulations (EC) No 854/2004 and (EC) No 882/2004 of the European Parliament and of the Council, Council Directives 89/608/EEC, 89/662/EEC, 90/425/EEC, 91/496/EEC, 96/23/EC, 96/93/EC and 97/78/EC and Council Decision 92/438/EEC (Official Controls Regulation) (Text with EEA relevance), OJ L 95, 7.4.2017.

[80] Directive 2004/48/EC of the European Parliament and of the Council of 29 April 2004 on the enforcement of intellectual property rights (Text with EEA relevance) OJ L 157, 30.4.2004.

authorities (CAs), and certain control tasks at producer level can be performed by control bodies (CBs). However, market controls (downstream) cannot be delegated to control bodies. This general structure varies from one Member State to another.

4.6.1 *Salame Felino*: Official Controls and Verification in Charge of the Member States

As said in *Salame Felino* case, national systems are still in place to ensure their administrative enforcement (i.e. official controls and verifications of compliance with product specifications established by producers and monitoring of the use of a GI in the market).[81]

EU GIs scheme would lose consistency if the producers were judge and party on its own verification and control. Citizens and consumers expect any GI and quality scheme to be backed up by a robust verification and control system, regardless of whether the product originates in the Union or a third country.[82]

Therefore, GIs are subject to the system of official controls, in accordance with the principles set out in Regulation (EU) 2017/625,[83] which includes a system of controls at all stages of production, processing and distribution.

Although the added value of GIs is based on consumer trust, the existence of a system of controls helps the producer to build this consumer trust. The GIs relies significantly on self-control, due diligence and individual responsibility of producers. There is a consistent message in Article 39 Regulation (EU) 2024/1143, where it is said that each operator should be subject to a control system that verifies compliance with the product specification.[84]

[81] Op. cit. 18, para. 28.

[82] Recital 9 of Regulation (EU) 2024/1143.

[83] See Article 26 Regulation (EU) 2017/625 of the European Parliament and of the Council of 15 March 2017 on official controls and other official activities performed to ensure the application of food and feed law, rules on animal health and welfare, plant health and plant protection products, amending Regulations (EC) No 999/2001, (EC) No 396/2005, (EC) No 1069/2009, (EC) No 1107/2009, (EU) No 1151/2012, (EU) No 652/2014, (EU) 2016/429 and (EU) 2016/2031 of the European Parliament and of the Council, Council Regulations (EC) No 1/2005 and (EC) No 1099/2009 and Council Directives 98/58/EC, 1999/74/EC, 2007/43/EC, 2008/119/EC and 2008/120/EC, and repealing Regulations (EC) No 854/2004 and (EC) No 882/2004 of the European Parliament and of the Council, Council Directives 89/608/EEC, 89/662/EEC, 90/425/EEC, 91/496/EEC, 96/23/EC, 96/93/EC and 97/78/EC and Council Decision 92/438/EEC (Official Controls Regulation) (OJ L 95, 7.4.2017, p. 1).

[84] Considering that wine is subject to specific controls laid down in Regulation (EU) No 1308/2013, Regulation (EU) 2024/1143 lays down controls for spirit drinks and agricultural products only.

Likewise, there will be one or more competent authorities[85] or delegated bodies or natural persons[86] to complement the verification of compliance with the product specification, in addition to the own controls and prior to place in the market a GI product and originating in the UE. To ensure that they are impartial and effective, the competent authorities designated to perform the verification of the compliance with the product specification must meet several operational criteria. In this sense, provisions on delegating certain official control tasks to delegated and product certification bodies and natural persons are recommended by the EU legislator to facilitate the task of the control authorities and make the system more effective. Information on the competent authorities, delegated and product certification bodies and natural persons should be made public to ensure transparency and allow interested parties to contact them.[87]

According to Article 38(2) Regulation (EU) 2024/1143 controls mean: (a) verification that a product designated by a GI has been produced in compliance with the corresponding product specification; and (b) verification of the use of GIs in the market, including on online interfaces.

Regarding the control and the verification of compliance with the product specification, Member States have several specific duties that are defined in Article 39 and 40 Regulation (EU) 2024/1143. First, they must draw up and keep up to date a list of operators who perform activities subject to one or more obligations provided for in the product specification of a product designated by a GI entered in the Union register of GIs originating in their territory. Second, they will make public the names and addresses of the competent authorities, delegated bodies and natural persons for each product designated by a GI and must keep that information up to date. In this context, Member States can collect fees or charges to cover, in whole or in part, the costs of official controls and other official activities.

---

[85] Article 3, point (3), of Regulation (EU) 2017/625.

[86] Title II, Chapter III, of Regulation (EU) 2017/625.

[87] European standards developed by the European Committee for Standardisation and international standards developed by the International Organisation for Standardisation should be used for the accreditation of the delegated and product certification bodies as well as by those bodies for their operations. The accreditation of those bodies should take place in accordance with Regulation (EC) No 765/2008 of the European Parliament and of the Council. According to Article 41 Regulation (EU) 2024/1143, the delegated bodies referred to in Article 39(3), point (b), and the product certification bodies referred to in Article 39(4), point (b), shall comply with and be accredited in accordance with either of the following standards as relevant for the delegated tasks: (a) Standard EN ISO/IEC 17065 'Conformity assessment — Requirements for bodies certifying products, processes and services'; or (b) Standard EN ISO/IEC 17020 'Conformity assessment — Requirements for the operation of various types of bodies performing inspection'. This accreditation referred to must be performed by a national accreditation body recognised in accordance with Regulation (EC) No 765/2008, that is a signatory of a Multilateral Agreement in the framework of the European Cooperation for Accreditation covering the standards referred to, or by an accreditation body outside the Union that is a signatory of a Multilateral Recognition Arrangement of the International Accreditation Forum or a Mutual Recognition Arrangement of the International Laboratory Accreditation Cooperation covering the standards referred to.

In the end, the EU GIs scheme is based on the consumer confidence, and this must be granted by a clean separation between the ones managing the GI and the ones controlling and verifying the GI. The role of the Commission is to audit the Member States based on a risk analysis.

4.6.2 The Adoption of the Necessary Measures to Deal with Conduct Which Infringes a PDO: *Commission v Denmark (PDO Feta)* and *Parmesan* Case

Under the EU GIs scheme, it corresponds to the Member States to define the necessary actions to tackle with conducts that infringes a PDO. In *Commission v Denmark (PDO Feta)*,[88] the Commission, supported by Greece and Cyprus, claimed that Denmark breached its obligations under Regulation (EU) No 1151/2012 by failing to prevent or stop the use of the designation 'Feta' on cheese produced in Denmark and intended for export to third countries. As quoted in chapter 1, when discussing on the internal market aspects of the EU GI scheme, the use of the PDO 'Feta' to designate products produced in the territory of the EU '*which do not comply with the product specification for that PDO undermines those objectives, even if those products are intended for export to third countries. It therefore follows from the wording of Regulation No 1151/2012, as well as from its context and the objectives pursued by it, that such use constitutes conduct prohibited by that regulation*'. The Court concluded that, by failing to prevent or stop such use in its territory, Denmark failed to fulfil its obligations under the EU GIs scheme.

In the *Parmesan* case,[89] the Commission requested the Court to declare that, by formally refusing to proceed against the use, on its territory, of the name 'Parmesan' on the labelling of products which did not comply with the requirements of the specification for the PDO 'Parmigiano Reggiano', thereby favouring the appropriation of the reputation of the genuine, Community-wide protected product, Germany failed to fulfil its obligations under the EU GIs scheme.[90] The German Government replied that, although the term 'Parmesan' had historical roots in the region of Parma, it had become a generic name for hard cheeses of diverse origins, grated or intended to be grated, distinct from the PDO 'Parmigiano Reggiano'.

In the case at hand, the Commission contends that the Federal Republic of Germany was bound, under Articles 10 and 13 of Regulation No 2081/92, to take on its own initiative the measures necessary to deal with conduct which infringes a PDO. According to the Commission, Member State intervention should include administrative and penal measures such as to enable the objectives referred to in that regulation concerning the protection of designations of origin to be achieved.

---

[88] C-159/20, Commission v Denmark (AOP Feta), ECLI:EU:C:2022:561, paras. 63–66.

[89] C-132/05—Commission v Germany, ECLI:EU:C:2008:117.

[90] At that time, the obligations listed in Article 13(1)(b) of Council Regulation (EEC) No 2081/92 of 14 July 1992 on the protection of geographical indications and designations of origin for agricultural products and foodstuffs (OJ 1992 L 208, p. 1).

Products which do not comply with the requirements of the regulation cannot be marketed.

The Commission pointed out that its complaints were directed not at the German legislation or at any lack of a right of action before the national courts, but at the administrative practice of the German authorities which was contrary to Community law. If the Member States were exempted from their obligation to intervene and if, consequently, economic operators themselves had to bring legal proceedings each time their exclusive right to use the PDO at issue throughout the territory of the European Union were infringed, the objectives of Regulation No 2081/92 could not be achieved. Again from the point of view of the Commission, the central question in legal proceedings between private economic operators was that of compliance with the intellectual property rights enjoyed by the producers established in the region of origin of the product concerned, whereas the purpose of action by the public authorities against infringements of Article 13 of Regulation No 2081/92 was not to protect private economic interests but those of consumers, whose expectations as to the quality and geographic origin of that product should not be disappointed. The protection of consumers intended by the regulation would be compromised if the enforcement of the prohibitions laid down by the regulation were completely dependent on the taking of legal action by private economic operators.[91] However, the Court concluded that there was not such an infringement and dismissed the action.

Regarding the Commission's complaint concerning the obligation on the Member States to take on their own initiative the measures necessary to penalise infringements, the Court made the following points:

> 73 First of all, there is no such obligation under Article 10 of Regulation No 2081/92.
>
> 74 It is true that, in order to ensure the effectiveness of the provisions of Regulation No 2081/92, Article 10(1) of that regulation provides that the Member States shall ensure that inspection structures are in place not later than six months after its entry into force. They are therefore obliged to create such structures.
>
> 75 Nevertheless, Article 10(4) of Regulation No 2081/92, by providing that '[i]f a designated inspection authority and/or private body in a Member State establishes that an agricultural product or a foodstuff bearing a protected name of origin in that Member State does not meet the criteria of the specification, they shall take the steps necessary to ensure that this Regulation is complied with ...', indicates that the designated inspection authority and/or private body in a Member State is that of the Member State from which the PDO comes.

Therefore, the Court pointed out that the obligation for the Member State consisted of granting that the inspection structures to be established in order to ensure the effectiveness of the EU GIs scheme. Therefore, the inspection structures must be created by the Member States.

[91] See op. cit. 87, paras. 60 and 61.

4.6.3 Enforcement and Online Intermediary Services: New Responsibilities for the Judicial or Administrative Authorities of the Member States Under the Digital Services Act

One of the pillars of the EU GIs system is that producers are properly rewarded for the added value of their products bearing a GI. In consequence, illegal users of those GIs are prevented from selling their products. To prevent fraudulent and deceptive practices and to allow the effective combating of counterfeiting, the enforcement of GIs in the market must be implemented in accordance with Regulation (EU) 2017/625, which contains specific rules on official controls and other official activities performed by the competent authorities in relation to GIs, as referred in the previous point related to official controls.

In the same vein, controls on the market should be carried out based on risk assessment or notifications from operators or competent authorities to ensure compliance with the product specification or single document or an equivalent to the latter, such as the summary of the product specification. With this aim, appropriate effective and proportionate administrative and judicial steps should be taken to prevent or stop the use of names on products or services that fail to respect, or contravene, the GIs in the market. The increased use of online intermediary services in relation to the enforcement of the protection of GIs against domains names that contravene that protection deserved the attention of the EU legislator who describes that it is necessary to equip the competent national authorities with the tools to react properly to a violation of the protection of a GI, by a registered domain name.[92]

In the view of the legislator, intermediary services, in particular online platforms, have become increasingly used for the sales of products, including those designated by GIs, and in some cases, they might represent an important space as regards preventing fraud.

Therefore, when exercising their official control tasks, those authorities should be able to take appropriate steps with a view to disabling access from the territory of the Member State concerned to domain names registered in breach of the protection of GIs, considering the principle of proportionality and the rights and interests of the affected parties. Such steps should be in line with other relevant Union law such as the Digital Services Act.[93]

Under Regulation (EU) 2024/1143 any information related to the advertising, promotion and sale of goods that contravenes the protection of GIs should be considered illegal content under the Digital Services Act. It is relevant, at this stage, to highlight that Article 3(h) of the Digital Services Act contains a definition of illegal content:

---

[92] See recital 55 Regulation (EU) 2024/1143.

[93] Regulation (EU) 2022/2065 of the European Parliament and of the Council of 19 October 2022 on a Single Market For Digital Services and amending Directive 2000/31/EC (Digital Services Act) (OJ L 277, 27.10.2022, p. 1).

> (h) 'illegal content' means any information that, in itself or in relation to an activity, including the sale of products or the provision of services, is not in compliance with Union law or the law of any Member State which is in compliance with Union law, irrespective of the precise subject matter or nature of that law;

From recital 12 of the Digital Service Act, it can be gathered that '*... for the purpose of this Regulation the concept of 'illegal content' should broadly reflect the existing rules in the offline environment. ... In particular, that concept should be understood to refer to information, irrespective of its form, that under the applicable law is either itself illegal, ... Illustrative examples include the sharing of images ... the sale of non-compliant or counterfeit products, the sale of products or the provision of services in infringement of consumer protection law,...*'.[94]

As a sort of *lex specialis*, Article 43 Regulation (EU) 2024/1143 provides for the identification of such illegal content as regards GIs:

> Any information related to the advertising, promotion and sale of products that is accessible to persons established in the Union and that contravenes the protection of geographical indications provided for in Articles 26 and 27 of this Regulation shall be considered illegal content as defined in Article 3, point (h), of Regulation (EU) 2022/2065.

Any person can alert online platforms about the presence of illegal content. Article 43(3) Regulation (EU) 2024/1143 declares that, in accordance with Article 16 of the Digital Services Act, any individual or entity may notify online platforms or providers of hosting services of the presence of a specific content that is in breach of Articles 26 and 27 Regulation (EU) 2024/1143. As it is known, both provisions contain the protection against any direct of indirect use of the GI; any misuse, imitation or evocation including when those products are used as an ingredient; any other false of misleading indication; or any other practice liable to mislead the consumer.

In any of these situations, the Member State judicial or administrative authorities are in charge of acting against the illegal content. In particular, the EU legislator stipulates that the relevant national judicial or administrative authorities of the Member States may, in accordance with Article 9 of the Digital Service Act, issue an order to act against the described illegal content affecting a GI; for instance, deciding on the geo-blocking of domain names that infringe GIs.

Likewise, the EU GIs scheme refers to the backbone of the Digital Services Act, in particular, Chapter III on 'Due diligence obligations'[95] as regards the online

[94] Recital 12 Regulation (EU) 2022/2065.

[95] Notification specifications only apply to reporting of manifestly *illegal* content. Thus, the violations of terms of use are left unregulated and subject to the providers' rules. If the notice specifications are fulfilled by notifiers (e.g., victims, NGOs, or authorities), this often (though not automatically) means that the providers are put on notice (i.e., if knowledge of the manifest illegality is relevant for their liability exemption, they acquired it and can lose exemption unless they act against that piece of content). See Husovec and Roche Laguna (2022), p. 6.

platforms and any other provider of hosting services; in particular the obligation of putting mechanisms in place to allow any individual or entity to notify them of the presence on their service of specific items of information that the individual or entity considers to be illegal content, known as notice handling.[96]

4.7 Conclusion

Our analysis of notable cases such as *Port Charlotte, Halloumi*, and *Comté* underscores the importance of effective verification, control, and enforcement mechanisms to ensure the authenticity and quality of GI products. We have highlighted the key procedural aspects of GIs, including registration, provisional protection, and national opposition procedures to develop conclusions on who does what in the EU GIs scheme for wine, spirit drinks and agricultural products. Following the referred landmark cases, like *Kühne* and *Jambon sec*, the general configuration of responsibilities reaches the logical conclusions that it is for the Member State to check the application; for the Commission to assess that the application contains the information required (i.e. standing orders) and it is not vitiated with errors; and for the National Courts to verify decisions taken by national authorities on minor amendments. The EU GIs scheme faces troubles when the national decision is reviewed by a national Court or when the Member State is called to adopt measures against conduct which infringe a PDO. As a matter of fact, *Commission v Denmark* (*PDO Feta*) and *Parmesan* Cases streamline that the Member states are in charge of the accuracy of the system of official controls, verification and inspection as well as when the products are in transit. To end up, the chapter answers the questions related to the new provisions affecting enforcement and online intermediary services contain and describes new responsibilities for the Member States that, for instance, will be required to geoblock domain names that may be infringing a GI.

References

Calabrese B (2024) The evolving protection of geographical indications against services: "brand" new world? IIC 55:348–367

Guillem Carrau, J (2021) Denominaciones geográficas de calidad. Estudio de su régimen jurídico en la OMC, la Unión europea y el Derecho español (Quality Geographical Indications. A study of the legal schem of the WTO, EU and Spanish law), Tirant lo blanch, 2 ed.

Guillem Carrau J (2022) How much protection for geographical indications (GIs)? A jigsaw of case law is falling into place in the EU. EIPR 44(9):556–566

Günzel O (2012) Opposing third parties' applications for geographical indications under Regulation 510/2006. J Intellect Prop Law Pract 7(4):244–250

Husovec M (2024) Principles of the Digital Services Act. Oxford

Husovec M, Roche Laguna I (2022) Digital Services Act: a short primer (July 5). https://ssrn.com/abstract=4153796

[96] Husovec (2024), p. 209.

Lobato García-Miján M (1995) La protección de las denominaciones de origen. Estudio del Reglamento (CEE) núm. 2081/1992 (The protection of the designations of origin. Study of Regulation (EEC) no 2081/1992), en Estudios de Derecho mercantil en homenaje al profesor Manuel Broseta Pont, VVAA, II, Valencia, Tirant lo Blanch, pp 1975 and 1993

Zappalaglio A (2015) The Piadina Romagnola Mess. A new legal case for an old question: what is a GI? (September 15). https://ssrn.com/abstract=2791400

5 The Singular Extension of Geographical Indications Protection as Intellectual Property Right

Abstract

The examination of the workings of GI protection is focused on the balance between origin and specificity, and the threat of genericity through a series of judgments, including *Feta*, *Grana Padano, Edam Holland*, and *Aceto Balsamico di Modena*. By exploring the scope of GI protection, the EU GC/CJ have delimitated the contours of the prohibition of direct and indirect use, misuse, and misleading practices, as seen in cases such as *Bavaria*, *Chianti,* and *Scotch Whisky*. Furthermore, it attempts to explain the criteria for assessing comparable products, as exemplified by the *Cognac* (BNIC) case, and the expansive protection against evocation, as demonstrated by *Manchego*, *Morbier* and *Champanillo*, without forgetting the protection under unfair competition law, reflected on *Salame Felino*.

One of the key challenges of the EU GIs scheme is striking a balance between protecting the intellectual property rights of producers and ensuring that the system is fair to all. This Chapter answers questions related to the examination of GIs and the balance between origin and specificity, and the threat of genericity through a series of key cases, including *Feta, Grana Padano, Edam Holland,* and *Aceto Balsamico di Modena*.

By exploring the real scope of GI protection, it attempts to describe how the EU General Court and Court of Justice explained the contours of the prohibition of direct and indirect use, misuse, and misleading practices, as seen in cases such as *Bavaria, Chianti*, and *Scotch Whisky Association.*

Furthermore, it seeks to provide an appropriate analysis of the criteria for assessing comparable products, as exemplified by the *Cognac (BNIC)* case, and the expansive protection against evocation, as demonstrated by *Manchego, Morbier* and *Champanillo*.

© The Author(s), under exclusive license to Springer Nature Switzerland AG 2026
J. Guillem Carrau, *Landmark EU Geographical Indications Case Law*, Law for Professionals, https://doi.org/10.1007/978-3-032-26101-4_5

Finally, it touches on the extension of GI protection under unfair competition law, reflecting on the *Salame Felino* case.

This examination attempts to draw conclusions on the extent of the GIs protection and to establish overall conclusions on the connection with the principles of, among others, veracity of origin, reputation, exclusivity and sustainability.

5.1 GI Protection Versus Genericity

Genericity was 'the Elephant in the *GI* Room'. Nobody wanted to discuss about the problem after the *Feta* cases saga.[1] In the very beginning, the identification of generic terms was done in cases where technical regulations affecting the internal market were assessed as not compatible with the internal market rules. We can quote, for instance, the following ones: *Prädikattsekt, Sekt and Weinbrand,*[2] *Jenever,*[3] *deep-frozen yogurt,*[4] *pasta (obligation to use durum wheat),*[5] *beer (purity requierement for),*[6] *edam,*[7] *vinegar,*[8] *cheese (prohibition on the marketing of cheeses not complying with Italian rules),*[9] *prepared meat product.*[10] They were terms for products that, once upon a time, were commercialised in a defined region and then become common, being produced abroad and in a significant volume and not fully respecting the original recipe. Currently, this examination is done in the context of Directive (EU) 2015/1535.[11]

A second set of landmark cases was on the definition of the concept of genericity and how to assess it. Thanks to judgments like the ones of the *Feta* saga, *Grana Biraghi, Edam Holland* and *Aceto Balsamico di Modena*, the EU legislator attempted the priority of clarifying the rules concerning the continued use of generic terms so that generic terms that are similar to or form part of a name or term that is protected must retain their generic status, as it is said in recital 37 Regulation (EU) 2024/1143.

> Generic term means the name of a product which, although relating to the place, region or country where a product was originally produced or placed on the market, has become the common name of a product in the Union.[12]

[1] C-289/96, C-293/96 y C-299/96, Feta ECLI:EU:C:1999:14, ECLI:EU:C:2005:636.

[2] C-12/74 - Commission v Germany, ECLI:EU:C:1975:23.

[3] C-182/84 – Miro. ECLI:EU:C:1985:470.

[4] C-298/87 – Smanor. ECLI:EU:C:1988:415.

[5] C-407/85 - 3 Glocken and Others v USL Centro-Sud and Others, ECLI:EU:C:1988:401.

[6] C-178/84 - Commission v Germany. ECLI:EU:C:1987:126.

[7] C-286/86 - Ministère public v Deserbais ECLI:EU:C:1988:434.

[8] C-193/80 - Commission v Italy ECLI:EU:C:1981:298.

[9] C-210/89 - Commission v Italy, ECLI:EU:C:1990:357.

[10] C-269/89 – Bonfait, ECLI:EU:C:1990:399.

[11] Directive (EU) 2015/1535 of the European Parliament and of the Council of 9 September 2015 laying down a procedure for the provision of information in the field of technical regulations and of rules on Information Society services, OJ L 241, 17.9.2015, pp. 1–15.

[12] Article 2.1 (h) Regulation (EU) 2024/1143.

Among all the relevant factors to establish whether a term has become generic, account shall be taken of the existing situation in the areas of consumption and the relevant national or Union legal acts. In addition, the EU legislator has clearly defined that GIs already registered must not become generic in the Union.

This constitutes an specific character that differentiates GI in the context of the intellectual property. It is said that we need to make the industrial property right owner liable for his right becoming generic and consider his actions or inactions. The right owner might use the term as generic or the relevant public might use the term as generic, and the right owner might do nothing to prevent that, even if he has the legal means to do so. Only with the ascription of liability to the right owner, can we consider that the term has become generic. Usually, the owner's behaviour is due to a lack of action when the term protected as an industrial property right starts to be used as a generic term by the relevant public (consumers of the concerned goods and trade opinion). To conclude, Article 28 Regulation (EU) 2024/1143 states that generic terms shall not be registered as a GI.[13]

5.1.1 *Feta saga:* Origin and Specificity Against Genericity

As it was previously explained, the *Feta* case is a helpful source to understand how a term considered generic in a certain moment can become protected under the GI system afterwards.[14]

As a matter of fact, in the case of Feta cheese, the reconsideration of the genericity relays on the elements that have been considered by the European Commission in the Commission Regulation (EC) No 1829/2002 of 14 October 2002 amending the Annex to Regulation (EC) No 1107/96 with regard to the name 'Feta'.

For the Court, it was fundamental the recital number 20 of the Commission Regulation (EC) No 1829/2002 with reference to Greek territory, culture or tradition done by Members States where *Feta* was produced at that moment:

> (20) According to the information sent by the Member States, those cheeses actually bearing the name "Feta" on Community territory generally make explicit or implicit reference to Greek territory, culture or tradition, even when produced in Member States other than Greece, by adding text or drawings with a marked Greek connotation. The link between the name "Feta" and Greece is thus deliberately suggested and sought as part of a sales strategy that capitalises on the reputation of the original product, and this creates a real risk of consumer confusion. Labels for "Feta" cheese not originating in Greece but actually marketed in the Community under that name without making any direct or indirect allusion to Greece are in the minority and the quantities of cheese actually marketed in this way account for a very small proportion of the Community market.

Likewise, the recital number 33 of Commission Regulation (EC) No 1829/2002, quoting the opinion of the Scientific Committee, was instrumental in dismissing the

[13] De Almeida and Carls (2021).

[14] Joined Cases C-465/02 and C-466/02, Denmark v Commission, Germany v Commission, ECLI:EU:C:2005:804 and ECLI:EU:C:2005:636.

notion that "Feta" was a generic term, nor a common name, for an agricultural product, as highlighted in the following paragraph:

> (33) The Commission has taken note of the advisory opinion of the Scientific Committee. It takes the view that the exhaustive overall analysis of the legal, historical, cultural, political, social, economic, scientific and technical information notified by the Member States or resulting from investigations undertaken or sponsored by the Commission leads to the conclusion that in particular none of the criteria required under Article 3 of Regulation (EEC) No 2081/92 to show that a name is generic have been met, and that consequently the name "Feta" has not become "the name of an agricultural product or a foodstuff which, although it relates to the place or the region where this product or foodstuff was originally produced or marketed, has become the common name of an agricultural product or a foodstuff".

It is also noteworthy that the Court quotes the recital number 35 of the above referred Regulation that contains the argument regarding specificity linked to the fact that *Feta* was a non-geographical name, with a unique characteristic being that its natural and human factors, which contribute to its reputation, were likely to extend beyond administrative borders.

> (35) The name "Feta" is a traditional non-geographical name within the meaning of Article 2(3) of Regulation (EEC) No 2081/92. The terms "region" and "place" mentioned in that provision may be interpreted only from a geomorphological and non-administrative viewpoint, in so far as the natural and human factors inherent in a given product are likely to transcend administrative borders. Under the above Article 2(3), however, the geographical area inherent in a designation may not cover an entire country. In the case of the name "Feta", it has therefore been noted that the defined geographical area referred to in the second indent of Article 2(2)(a) of Regulation (EEC) No 2081/92 covers only the territory of mainland Greece and the department of Lesbos; all other islands and archipelagos are excluded because the necessary natural and/or human factors do not apply there. Moreover, the administrative definition of the geographical area has been refined and developed, since the product specification submitted by the Greek authorities contains mandatory cumulative requirements: in particular, the area of origin of the raw material has been substantially limited since the milk used to produce "Feta" cheese must come from ewes and goats of local breeds reared traditionally, whose feed must be based on the flora present in the pastures of eligible regions.

Lastly, another key recital of the Commission Regulation (EC) No 1829/2002 quoted by the Court is the one related to the consistency of the geographical area governed by the administrative decision, and the dynamic interaction between natural factors and specific human factors that influence the product's reputation and characteristics.

> (36) The geographical area covered by the administrative definition and meeting the requirements of the product specification is sufficiently uniform to meet the requirements of Articles 2(2)(a) and 4(2)(f) of Regulation (EEC) No 2081/92. Extensive grazing and transhumance, central to the method of keeping the ewes and goats used to provide the raw material for making "Feta" cheese, are the result of an ancestral tradition allowing adaptation to climate changes and their impact on the available vegetation. This has led to the development of small native breeds of sheep and goats which are extremely tough and resistant, fitted for survival in an environment that offers little food in quantitative terms but, in terms of quality, is endowed with an extremely diversified flora, thus giving the finished

> product its own specific aroma and flavour. The interplay between the above natural factors and the specific human factors, in particular the traditional production method, which requires straining without pressure, has thus given "Feta" cheese its remarkable international reputation.

For the Court, since the term "Feta" had not been established as generic, the Commission correctly verified that the application by the Greek authorities for the name "Feta" complied with the requirements to be protected as PDO.

In a nutshell, when considering the registration of *Feta PDO*, the analysis of the legal, historical, cultural, political, social, economic, scientific and technical information must draw up a picture of the natural factors and the specific human ones that makes the remarkable reputation of a product to be under the protection of the EU GIs scheme.

5.1.2 Grana Padano: The Generic Character Is Not About the Definitions in a Dictionary But About Market Share and Legislative Positions Among Other Factors

If we go back to *Grana Biraghi*,[15] it provides us with a detailed summary of the criteria adopted to assess genericness. In conclusion, we find that, in the view of the Court, when determining if a term has become generic, it is essential to look at a serial of factors like the relevant economic, social and historic evidences as well as the legislation, the market share held by the protected goods in comparison with the non-protected ones, the way of presenting the protected products, the protection under international agreements, etc.:

> 65. Thus, the legal, economic, technical, historical, cultural and social evidence which makes it possible to carry out the necessary detailed analysis is, inter alia, the relevant national and Community legislation, including its historical development, the perception which the average consumer has of the allegedly generic name, including the fact that the reputation of the name remains linked to the traditional matured cheese produced in a rural area as a result of the fact that it is not commonly used in other areas of the Member State or of the European Union, the fact that a product has been legally marketed under the name in question in certain Member States, the fact that a product has been legally produced under the name in question in the country where the name originated even though the traditional methods for the production thereof have not been complied with, the fact that such processes have endured over time, the quantity of goods which bear the name in question and are produced using non-traditional methods as against the quantity of goods produced using traditional methods, the market share held by goods bearing the name in question which were produced using non-traditional methods as against the market share held by goods produced using traditional methods, the fact that the goods produced using non-traditional methods are presented in such a way as to refer to the places of production of goods produced using traditional methods, the protection of the name in question under international agreements and the number of Member States which may rely on the allegedly generic nature of the name in question.

[15] T-291/03, Grana Padano, EU:T:2007:255, para. 65–66.

Moreover, the results of a survey of consumers can constitute also a factor to be considered, as highlighted by the Court in the case at hand:

> 66 Furthermore, the Court has not excluded the possibility of taking account (Denmark and Others v Commission, paragraphs 85 to 87), in an examination of the generic nature of a name, of a survey of consumers organised in order to understand their perception of the name in question or of an opinion of the committee set up by Commission Decision 93/53/EEC of 21 December 1992 setting up a scientific committee for designations of origin, geographical indications and certificates of specific character (OJ 1993 L 13, p. 16), which has since been replaced by the scientific group of experts for designations of origin, geographical indications and traditional specialities guaranteed, set up by Commission Decision 2007/71/EC of 20 December 2006 (OJ 2007 L 32, p. 177). That committee, made up of highly qualified experts in the fields of law and agriculture, has the task of examining, inter alia, the generic nature of names.

At this stage, other evidence as the definition of the term in the *Codex Alimentarius* or the *Stressa* Convention[16] can be reflected in the assessment of the genericity of the term:

> 67 Lastly, it is possible to take into consideration other factors, inter alia, the definition of a name as generic in the Codex alimentarius (on the indicative value of the rules of the Codex alimentarius see Case 286/86 Deserbais [1988] ECR 4907, paragraph 15, and Case C-448/98 Guimont [2000] ECR I-10663, paragraph 32) and the inclusion of the name on the list in Annex II to the international Convention for the use of designations of origin and names of cheeses, signed at Stresa on 1 June 1951, in that that inclusion permits the use of the name in any country signatory to the Convention, provided that the rules of manufacture are complied with and that the country of production is mentioned, without restricting that use to manufacturers in the corresponding geographical area (see the Opinion of Advocate General Ruiz-Jarabo Colomer in Germany and Denmark v Commission, point 168).

Likewise, in *Grana Biraghi,* the discussion was about the presence of GI terms in dictionaries and the internet and its impact of this fact in the assessment of the genericity. In the case at hand, for instance, the parties argued that the term ‘granas’ was defined in the *Dictionnaire des fromages Larousse* as ‘Italian cheeses which have (some) common characteristics’.

This was not relevant at all for the Court because a number of Italian laws had, at that time, defined the designation by distinguishing the cheeses produced in certain provinces (parmigiano reggiano) from those produced in certain other provinces (grana padano). All this was in support of the argument that the name ‘grana’ was not generic in nature. It was apparent from the evidence examined that the Board of

---

[16] This Convention, whose entry in force was on 1953 recognised a level of protection to the following geographical terms as PDO for cheese products: *Gorgonzola, Parmigiano Reggiano, Pecorino Romano and Roquefort;* and a second level to a large list of terms like *Camembert, Danablu, Edam, Emmental,* etc. The compatibility of this Convention with the free circulation of goods was discussed in *Deserbais y Cambozola* [C-286/86 - Ministère public v Deserbais, ECLI:EU:C:1988:434, and C-87/97 - Consorzio per la tutela del formaggio Gorgonzola, ECLI:EU:C:1999:115]. See Guillem Carrau (2021), p. 74.

Appeal that registration of the mark GRANA BIRAGHI did constitute an infringement of the PDO 'Grana Padano'.

As said, the evidence substantiating the contested decision consisted merely of extracts from dictionaries and internet research carried out of its own motion by the Board of Appeal:

> 71 The frequency with which a term appears on the internet is not, of itself, capable of establishing the generic nature of a name. Furthermore, all the definitions of the term 'grana' given in the dictionaries cited by the Board of Appeal refer to the place of production of grana padano, which is an area in the plain of the Po. Consequently, and contrary to the findings of the Board of Appeal, those dictionaries show that the name 'grana' is used in Italian as an abbreviated form of grana padano and that the name 'grana' is linked in fact and in people's minds with the Padanian origin of that product, which is borne out by the two German dictionaries referred to in paragraphs 50 and 51 of the contested decision. As regards the definition given in the Enciclopedia Treccani, that definition is irrelevant in so far as it dates from 1949, that is to say from a date prior to both Regulation No 1107/96 and Law No 125/54, which first acknowledged the name 'grana padano' as a PDO.

As referred in chapter 2, in *Grana Biraghi*, the Court, among those factors, mention made, above all, of the legislative position as regards the protection of the name 'grana padano' in Italy and the historical development thereof.[17]

5.1.3 *Edam Holland* and *Aceto Balsamico di Modena*: A Protection Which Does Not Cover Generic Element

Where a PDO or a PGI is made up of several elements/names, one or more of which is the name of a product which is considered generic, the use by a third party of the generic element/name does not, in principle, infringe the protection granted against any direct or indirect commercial use of a registered name and any misuse, imitation or evocation of a registered name respectively.

This approach was confirmed by the Court in *Edam Holland*.[18] The term *Edam*, as such, can be used after the registration of *Edam Holland*:

> 21. Recital 8 in the preamble to the contested Regulation states, by expressly referencing Article 13(1), second subparagraph, of the basic Regulation, that 'the term "Edam" may continue to be used'. That provision of the basic Regulation provided that, when the name of an agricultural product or foodstuff, which is part of a name registered in the register of protected designations of origin and protected geographical indications, is generic, the commercial use of that name does not infringe the protection of the aforementioned registered name. In this respect the Commission determined, in Article 1 of the contested Regulation, that, since it constitutes a generic designation, the name 'Edam' could, notwithstanding the registration of the PGI at issue, continue to be used within the territory of the European Union, provided that the principles and rules applicable in its legal order were respected.

[17] See op. cit 15, para. 73.

[18] C-517/14 P, Schutzgemeinschaft Milch und Milcherzeugnisse v Commission, EU:C:2015:700, para. 21.

> 22 It follows that the General Court did not err in law in finding, in paragraphs 29 and 30 of the order under appeal, that annulment of the contested Regulation would not be of any advantage to the appellant's members since that regulation provides that the name 'Edam' may continue to be used, inter alia for the marketing of cheeses, on condition that the principles and rules applicable in the European Union's legal order are respected.

In *Aceto Balsamico di Modena,*[19] the GC refers that protection covers the whole term and it does not extend to the use of the non-geographical terms of that name such as 'aceto' and 'balsamico'.

> 33 It thus follows unequivocally from the recitals of Regulation No 583/2009 that the non-geographical terms of the PGI at issue, namely 'aceto' and 'balsamico', and their use in combination and in translation cannot benefit from the protection for the PGI 'Aceto Balsamico di Modena' that was granted by Regulation No 510/2006 and is now ensured by Regulation No 1151/2012.
>
> 34 Moreover, first, it is established that the term 'aceto' is a common term, as previously held by the Court (see, to that effect, judgment of 9 December 1981, Commission v Italy, 193/80, EU:C:1981:298, paragraphs 25 and 26). Second, the term 'balsamico' is the Italian translation of the adjective 'balsamic' which has no geographical connotation and which, in the case of vinegar, is commonly used to refer to a vinegar with a bitter-sweet flavour. It is therefore also a common term within the meaning of the case-law referred to in paragraph 26 of this judgment.

Moreover, the GC confirmed the crucial role of the European Commission in the process of registration because the Act registering the GI already stated that the protection is granted to the term as such, so the non-geographical components of the term may be used, even jointly and also in translation, throughout the EU.

> 35 Lastly, as the Advocate General has also pointed out, in essence, in points 57 and 58 of his Opinion, that interpretation of the scope of the protection conferred on the PGI at issue is appropriate in the light of the registrations of the PDOs 'Aceto balsamico tradizionale di Modena' and 'Aceto balsamico tradizionale di Reggio Emilia' which were, moreover, as stated in the recitals to Regulation No 583/2009, also taken into account by the Commission when that regulation was adopted. The use in the text of those PDOs of the terms 'aceto' and 'balsamico' and their use in combination and in translation cannot be considered likely to infringe the protection conferred on the PGI at issue.

According to case law, the exclusive use is limited to the protected term and not to the generic parts of the compound name/term that are not expressly protected.[20]

---

[19] C-432/18, Consorzio Tutela Aceto Balsamico di Modena. ECLI:EU:C:2019:1045, para 34.

[20] Op. cit. 15, para. 58; T-828/16, Torta del Casar, ECLI:EU:T:2017:918; T-237/08, Cuvée Palomar, ECLI:EU:T:2010:185. See also Article 26(7) Regulation 2024/1143.

5.2 The Extent of GIs Protection: A Jigsaw of Case Law Is Falling into Place

The EU legislator has granted to GIs a broad scope of protection, which is less constrained by means of burdens of proof and limitations than trademarks. On top of that, the CJEU begun in 1997 a case-specific race to the bottom which have resulted on an expanded scope of protection of GIs against evocation. This is not a confirmation bias based on looking for or overvaluing case law that confirms GIs development expectations. The results of the case-specific race of the Court are visible in the text of Article 26 Regulation (EU) 2024/1143, which states that GIs entered in the Union register of GIs shall be protected against:

(a) *any direct or indirect commercial use of the geographical indication in respect of products not covered by the registration, where those products are comparable to the products registered under that name or where use of that geographical indication for any product or any service exploits, weakens, dilutes, or is detrimental to the reputation of, the protected name, including when those products are used as an ingredient;*
(b) *any misuse, imitation or evocation, even if the true origin of the products or services is indicated or if the protected name is translated, transcribed or transliterated or accompanied by an expression such as 'style', 'type', 'method', 'as produced in', 'imitation', 'flavour', 'like' or similar, including when those products are used as an ingredient;*
(c) *any other false or misleading indication as to the provenance, origin, nature or essential qualities of the product that is used on the inner or outer packaging, on advertising material, in documents or information provided on online interfaces relating to the product concerned, and the packing of the product in a container liable to convey a false impression as to its origin;*
(d) *any other practice liable to mislead the consumer as to the true origin of the product.*

As it is said in *Scotch Whisky Association*[21] and in *Morbier*[22] the above situations work as a kind of '*graduated list of prohibited conducts*'. Where the first indent refers to the use of the name itself; the second one tackle the use of figurative signs, or other type of elements, capable of evoking in the mind of the consumer products whose designation is protected; the third one further widens the scope of protection to any false or misleading information provided to consumers, on the inner or outer packaging of the product concerned, on advertising material or documents relating to that product, even if not evocative of the indication; and the forth one simply intends to catch any conduct not covered by the other provisions.[23]

---

[21] C-44/17, Scotch Whisky Association, ECLI:EU:C:2018:415, para. 33.

[22] C 490/19, Syndicat interprofessionnel de défense du fromage Morbier, EU:C:2020:1043, para. 24.

[23] Günzel (2021), p. 294.

The expression '*graduate list of prohibited conducts*' is also recited in *Champanillo*[24] and in *Portsoy* cases.[25] In fact, this expanded scope is not comparable to any other EU IP right, nor to GI protection in other jurisdictions[26]

The impact of the case law is fully visible in the wording of Article 26 Regulation (EU) 2024/1143 by comparison with the previous provision.[27] Regarding the '*any direct or indirect commercial use*' of the GI, the insertion of '*weakens, dilutes, or is detrimental to the reputation of*' instead of '*exploits the reputation of*' will increase the protection of the GI and it is consistent with the case law that is going to be explained later on. Likewise, as to the '*any misuse, imitation or evocation*' the insertion of '*transcribed or transliterated*' will be equally redundant in a better protection of the registered GIs.

However, the type of protection that a GIs enjoys is not absolute. One of the first limits that encounter the protection is the pre-existence of homonymous terms. It is forbidden to register homonymous titles unless there is sufficient distinction in practice between the conditions of local and long-established usage and the presentation of the two wholly or partially homonymous indications, considering the need to ensure equitable treatment of the producers concerned and that consumers are not misled as to the identity or geographical origin of the products.

A wholly or partially homonymous GI which misleads the consumer into believing that products come from another territory must not be registered even if the name for the actual territory, region, or place of origin of the products concerned is accurate. This could happen in several situations, as the legislator describes: (a) GIs already registered; (b) GIs applied for provided that they are subsequently registered; (c) those protected in the Union in accordance with Regulation (EU) 2019/1753; and (d) GIs, names of origin and equivalent terms protected pursuant to an international agreement between the Union and one or more third countries.[28]

---

[24] C-783/19, Comité Interprofessionnel du Vin de Champagne, EU:C:2021:713, para. 36.

[25] T-40/24 - IVDP v EUIPO - The Benriach Distillery Company (PORTSOY) ECLI:EU:T:2025:183, para 19.

[26] Zafrilla Díaz-Marta and Kyrylenko (2021), p. 449.

[27] See the previous version of Article 26 Regulation (EU) 2024/1143 in Article 13 Regulation (EU) 1151/2012 which said '*registered names shall be protected against:*

(a) any direct or indirect commercial use of a registered name in respect of products not covered by the registration where those products are comparable to the products registered under that name or where using the name exploits the reputation of the protected name, including when those products are used as an ingredient;

(b) any misuse, imitation or evocation, even if the true origin of the products or services is indicated or if the protected name is translated or accompanied by an expression such as 'style', 'type', 'method', 'as produced in', 'imitation' or similar, including when those products are used as an ingredient;

(c) any other false or misleading indication as to the provenance, origin, nature or essential qualities of the product that is used on the inner or outer packaging, advertising material or documents relating to the product concerned, and the packing of the product in a container liable to convey a false impression as to its origin;

(d) any other practice liable to mislead the consumer as to the true origin of the product.'

[28] See Article 29 Regulation (EU) 2024/1143.

5.2.1 A Protection Against Any Direct or Indirect Use and Against Misuse: *Chianti, Cuvée Palomar, Riojavina, Cambozola*

The EU GIs scheme seeks to prohibit any direct or indirect use of a registered name exploiting the reputation of a GI protected by registration, in a form that is phonetically and visually identical with that name or, at least, highly similar, as stated by the Court in *Morbier*.[29]

According to the case law, the direct or indirect use happens when there is a high degree of visual and aural similarity and the products are identical or there is a commercially relevant ingredient, or they are comparable.[30] By incorporating in the name of the foodstuff in question the name of the ingredient protected by a PDO, direct use is made of the PDO to claim openly a gustatory quality connected with it, which does not amount to misuse, imitation or evocation.[31]

In other words, the legislator forbids not use the protected name itself either directly or indirectly but suggest it in such a way that it causes the consumer to establish a sufficiently close connection with that name. In *Scotch Whisky Association* and *Morbier*,[32] the Court has clarified that a distinction must be drawn between situations in which the use is 'direct' and those in which it is 'indirect'. More specifically, there is 'indirect commercial use' of a PGI, when the disputed element is used in a form that is 'either identical to that indication or phonetically and/or visually similar to it'.[33]

> 39 In view of all the foregoing considerations, the answer to the first question is that Article 16(a) of Regulation No 110/2008 must be interpreted as meaning that, for the purpose of establishing that there is 'indirect commercial use' of a registered geographical indication, the disputed element must be used in a form that is either identical to that indication or phonetically and/or visually similar to it. Accordingly, it is not sufficient that that element is liable to evoke in the relevant public some kind of association with the indication concerned or the geographical area relating thereto.

This is particularly clearly stated in the Advocate General's opinion: "unlike 'direct' use, which implies that the PDO is affixed directly to the products in question or their packaging, 'indirect' use requires the indication to feature in supplementary marketing or information sources, such as an advertisement. or documents relating to it".[34]

As the Advocate General recalled in *Champanillo*, the wording of Article 103(2)(a)(ii), by referring to 'any direct or indirect commercial use', prevents its scope from being confined to the use of products only, to the exclusion of services and

[29] Op. cit.22 para. 24.

[30] C-393/16, Champagne Sorbet ECLI:EU:C:2017:991, para. 36.

[31] Op. cit. 25, para. 57.

[32] Op. cit 22, para. 33, and Op. cit.22 para. 25.

[33] Op cit 22, para. 39.

[34] AG Opinion Case C-44/77 Scotch Whisky Association ECLI:EU:C:2018:111, para. 30.

'Moreover, in specifying that the prohibition applies 'in so far as such use exploits the reputation of a designation of origin or a geographical indication', that provision specifically emphasises the effects of the use, and not a particular type of use. Second, the same considerations regarding the objectives of Regulation No 1308/2013 and the scope of protection afforded by that regulation to geographical indications registered in accordance with its provisions, set out in points 35 and 36 of this Opinion, also apply for the purposes of a teleological interpretation of Article 103(2)(a)(ii) of that regulation'.[35] In the case at hand, like its Advocate General, the Court examined, first, whether use of the sign '*Champanillo*' constituted a commercial 'use' of the PDO, rather than an 'evocation' thereof. It recalled that for a situation to constitute 'use' of a PDO or a PGI, the disputed sign must use the PDO or PGI 'in an identical form or at least in a form that is phonetically and/or visually highly similar'.[36] The Court held that this was not the case here in *Champanillo*[37]:

> 39. Unlike the conduct referred to in Article 103(2)(a) of Regulation No 1308/2013, the actions which fall within the scope of Article 103(2)(b) thereof do not use the protected name itself either directly or indirectly, but suggest it in such a way that it causes the consumer to establish a sufficiently close connection with that name (see, by analogy, judgments of 7 June 2018, Scotch Whisky Association, C-44/17, EU:C:2018:415, paragraph 33, and of 17 December 2020, Syndicat interprofessionnel de défense du fromage Morbier, C-490/19, EU:C:2020:1043, paragraph 25).
>
> 40 It follows that the concept of 'use', within the meaning of Article 103(2)(a) of Regulation No 1308/2013, must be interpreted strictly, as otherwise the distinction between that concept and, in particular, that of 'evocation', within the meaning of Article 103(2)(b) of that regulation, would be rendered moot, which would be contrary to the intention of the EU legislature.

Endorsing the Advocate General's Opinion, the Court pointed out that an excessively broad interpretation of the concept of 'use', including in situations where the alleged unlawful practice consists of the use of the PDO in a form that does not substantially correspond to the name as registered, is likely to blur the boundaries between the cases covered by Article 103(2)(a) of the Wine Regulation (dealing with 'uses' of a PDO) and those within the scope of Article 103(2)(b) (dealing with 'misuses, imitations and evocations'), making it hard to distinguish between them.[38]

Another relevant case is *Chianti*,[39] where the direct use of a rooster represented in a similar way was under discussion. The Court concluded the existence of a certain degree of visual similarity, by reason of the presence, in the centre of the marks, of a rooster represented in a similar way, as the Board of Appeal correctly emphasised in paragraph 20 of the contested decision, was weakened by the phonetic and conceptual differences, to the extent that

---

[35] AG Opinion C-783/19, Comité Interprofessionnel du Vin de Champagne v GB ECLI:EU:C:2021:350.

[36] Op. cit 18, para 24.

[37] Op. cit 25, para. 40.

[38] Vrins (2021), p. 1171.

[39] T-143/11 - Consorzio vino Chianti Classico v OHMI - FFR (F.F.R.), EU:T:2012:645.

it must be found that there was a low degree of overall similarity between the marks and it was unlikely that the relevant public could believe that the goods in question come from the same undertaking or from economically-linked undertakings:

> 58 So far as concerns the alleged distinctive character of the earlier marks, the Board of Appeal held, in paragraphs 25 and 26 of the contested decision, that the applicant had claimed and established the public's recognition of the single figure of a 'black rooster', associated with Chianti Classico wines, without making any reference to, in particular, the enhanced distinctive character of an earlier mark. The applicant does not dispute that finding. Moreover, as the Board of Appeal correctly held, the figure of the black rooster is not reproduced in the mark applied for, which contains an orange rooster. Therefore, the relevant public will not associate figures in very different colours, such as the striking orange colour used in the mark applied for, with those wines.
>
> 59 Thus, even though the earlier marks have enhanced distinctive character, that fact cannot compensate for the absence of a likelihood of confusion between the signs at issue within the meaning of Article 8(1)(b) of Regulation No 207/2009, in view of the conceptual, phonetic and visual differences, as well as the significance to be attributed, in the present case, to the differences between the word elements of the marks at issue.
>
> 60 In the light of the foregoing, it must be found that it is unlikely that the relevant public could believe that the goods in question come from the same undertaking or from economically-linked undertakings. Accordingly, the Board of Appeal acted correctly in law in holding that there was no likelihood of confusion between the marks at issue within the meaning of Article 8(1)(b) of Regulation No 207/2009.

In *Cuvée Palomar*, the Court endorsed the Boards of Appeal criteria on the similarity between el Palomar, which is the name of a local administrative area in the sub-region Clariano and constitutes, an area of production protected by the PDO 'valencia' and the requested EUTM. The Board of Appeal considered that its use is therefore prohibited pursuant to Article 2(2) of the Ministerial Order of 19 October 2000, since the Community trade mark had not been sought to designate wines from that area of origin.[40]

In *Riojavina*, the Court confirmed in essence that the similarity assessment done by the Boards of Appeal, which was basically to conclude the similarity, firstly because of the dominant denominative element 'Rioja', in common in both signs, as from their positions in the signs as from the reputation, not controversial between the parties, in a great part of the territory of the Union in relation to Rioja wines.[41]

The Court clearly stated in *Gorgonzola* that it was for the national court to determine whether the trade mark concerned has been registered in good faith and whether it was liable to be declared invalid or revoked on the basis of the Trade Marks Directive.[42]

To conclude with another aligned not appealed decision of the Boards of Appeal; in *Colombiano Coffe House*,[43] the Board of Appeal set up that misuse of the GI could happen when using the GI in an inappropriate context, which would be detri-

[40] T-237/08 - Abadía Retuerta v OHIM (CUVÉE PALOMAR), EU:T:2010:185, para. 23–25.

[41] T-138/09, Riojavina, EU:T:2010:226.

[42] Op. cit 16, para. 36–42.

[43] 30/11/2018, R 251/2016-1, Colombiano Coffee House (fig.).

mental to the reputation of quality products (e.g. Café de Colombia, for industrial paintings). In this case, as describe in chapter 7 of this book, the Board of Appeal applied the *BNIC*[44] criteria and found that that the goods in Class 30, apart from tea and cocoa, could not be comparable to coffee.

5.2.2 A Protection Against Any Other False or Misleading Indication as to the Origin: *Scotch Whisky Association* Highlights

Misleading the consumer means any other false or misleading indication as to the provenance, origin, nature or essential qualities of the product on the inner or outer packaging, advertising material or documents relating to the product concerned, and the packaging of the product in a container liable to convey a false impression as to its origin and any other practice liable to mislead the consumer as to the true origin of the product. It is possible to join the opinion expressed in literature that the broad protection described leaves little space for the rest of provisions, which in turn refer to questions when there is a chance to mislead consumers in circumstances when the protection against direct and indirect use cannot be applied as in case of homonymous indication or untrue packaging. Intellectual property protection is characterized as absolutely irrespective of misleading consumers; this principle is also known in trade mark law and refers only to the marks that are identical and have been used or registered for identical goods.[45]

In *Scotch Whisky Association*,[46] the Landgericht Hamburg (Regional Court, Hamburg) raised several questions in a preliminary ruling on the extension of the GIs protection. It was in the context of an action seeking an order that a stakeholder cease to market a whisky produced in Germany under the designation 'Glen Buchenbach' as it was a play on words. Among other questions, the Regional Court asked about when determining whether there is any "other false or misleading indication" in accordance with Article 16(c) of [Regulation No 110/2008]; in particular, whether the disputed element is embedded play a role, or can that context not counteract any misleading indication, even if the disputed element is accompanied by an indication of the true origin of the product.

The GIs producers argued in the case at hand that the designation 'Glen' was very widely used in Scotland instead of the word 'valley' and, in particular, as an element of the trade mark in the names of Scottish whiskies, it evoked in the relevant public an association with Scotland and Scotch Whisky despite the fact that other information was included on the label, which specified that the product at issue is of German origin. For the Court, it was not relevant to take account either of the context surrounding the disputed element, or, in particular, of the fact that that

---

[44] C-4/10 and C-27/10, BNIC, EU:C:2011:484, para. 54.

[45] Mantrov (2012), p. 193.

[46] Op. cit 21, para. 47–56.

element was accompanied by an indication of the true origin of the product concerned, or of the context in which the disputed element was used.

In that regard, the national court must, in making that assessment, refer to the perception of an average European consumer who is reasonably well informed and reasonably observant and circumspect.[47]

5.2.3 *Cognac (BNIC)*: The Criteria to Assess Comparable Products

In *Cognac*,[48] also known as *BNIC*, the Korkein hallinto-oikeus (Supreme Administrative Court) raised a preliminary ruling to refer some questions on the extension of the GI protection when the products are comparable as it was in the case at hand where two mark applications contained the terms 'Konjakit' and 'KONJAKKI Café Cognac' for goods in Class 33. By answering the questions, the Court developed the criteria for determining whether products are comparable or not. In particular, the Court identified the following non-exclusive list of factors:

a) Whether the products have common objective characteristics, such as method of elaboration, physical appearance or use of the same material;
b) Which are the circumstances under which the products are consumed, from the point of view of the relevant public;
c) Whether they are frequently distributed through the same channels; and
d) Whether they subject to similar marketing rules.

Comparability of products is assessed from the perspective of the average consumer. From *Gorgonzola* to *Portsoy*, passing by *Viniiverla* and *Scotch Whisky Association,* the Court reveals that it is an average European consumer who is reasonably well informed and reasonably observant and circumspect that concept being understood as covering European consumers and not only consumers of the Member State in which the product giving rise to the use or evocation of the PDO is manufactured.[49]

However, in *Manchego cheese*[50] case, the Court said that it must be understood as covering European consumers, including consumers of the Member State in which the product giving rise to evocation of the protected name is made or with which that name is geographically associated and in which the product is mainly consumed. Accordingly, a conflict assessed by reference to the consumers of a single Member State is sufficient to trigger the protection provided for by the EU regulations.

[47] C-75/15, Viiniverla, EU:C:2016:35, para. 25 and 28.

[48] Op. Cit 44, para. 54.

[49] See C-87/97, Gorgonzola, EU:C:1999:115, para. 25; Op. cit. 21, para. 47; Op. cit 25, para. 21.

[50] C-614/17, Fundación Consejo Regulador de la Denominación de Origen Protegida Queso Manchego, ECLI:EU:C:2019:344, para. 47 to 50.

Nevertheless, the issue of whether the products are identical or comparable must be taken into consideration to assess, when examining all the relevant circumstances, whether the conditions for evocation are in fact satisfied. The fact that those products have objective characteristics in common, that they are consumed on occasions which are largely identical, or have a similar appearance, and also that they are in competition with each other or are complementary, is therefore relevant to the assessment, as is the fact that, where the disputed sign relates to a service, this is linked to the distribution of the product covered by the registered name or of an identical or comparable product.[51]

Therefore, the scope of protection of a designation of origin makes it possible to protect that designation of origin not only as against comparable products but also as against any services which may be associated with the direct or indirect distribution of those products.[52]

One of the aspects that has been reviewed in the revision of the system performed with Regulation (EU) 2024/1143 is the concept of comparability. Firstly, by making crystal clear that the classification of the products according with the Combined Nomenclature shall not be used to determine comparable products for the purposes of protection against direct and indirect commercial use as referred to in Article 26(1), point (a).[53] Secondly, when describing that the GIs shall be protected against the direct or indirect use in comparable products not covered or when the use for any product exploits the GI reputation, the legislator added to these situations the following ones 'weakens, dilutes, or is detrimental to the reputation'. Recital 34 of Regulation (EU) 2024/1143, refers the *BNIC* criteria for comparability of products by saying that, whether products are comparable to the products designated by a GI, account should be taken of all relevant factors. As described before, those factors should include whether the products have common objective characteristics, such as method of production, physical appearance, or use of the same raw material; under which circumstances the products are used from the point of view of the relevant public; whether they are frequently distributed through the same channels; and whether they are subject to similar marketing rules. Therefore, there would not be a specific impact as regards the assessment of comparability. Thirdly, the Commission is empowered to adopt additional rules on the use of comparable products as ingredients and the criteria of conferring essential characteristics on the processed products.

5.2.4 An Expansive Protection Against Evocation: *Scotch Whisky, Manchego* and *Champanillo*

Building on the established case-law of the Court of Justice of the European Union, evocation of a GI may arise where a link with the product designated by the GI,

[51] Op. cit 35, para. 56.

[52] The Court endorsed AG Opinion in Champanillo case. Op. cit 24.

[53] See Article 6 Regulation (EU) 2024/1143.

including with reference to a term, sign, or other labelling or packaging device, is present in the mind of the average European consumer who is reasonably well-informed, observant and circumspect.[54]

In *Scotch Whisky Association* and in *Manchego cheese*, the Court has developed 'expansive' arguments as regards 'evocation'. Evocation can be detected even by figurative signs reproducing any graphic element of the GI's logos.[55] The concept of 'evocation' covers a situation in which the term used to designate a product incorporates part of a protected designation, so that when the consumer is confronted with the name of the product the image triggered in his mind is that of the product whose designation is protected. There is evocation of a protected name where, concerning products which are similar in appearance, the sales names are phonetically and visually similar even if the true origin of the product is indicated as happened with *Cambozola* as regards PDO *Gorgonzola* or with *Verlados* as regards PDO *Calvados*.[56]

In *Scotch Whisky Association*, as referred before, a company marketed whisky produced in Germany under the designation 'Glen Buchenbach', infringing the GI 'Scotch Whisky'. The GC clarifies the extension of the rights of the GIs tenants against other stakeholders that can try to benefit from the presence of the GI in the market. In particular, the case was about the use of the term 'Glenn', which is used in Scotland and Ireland, to describe a valley, in the label of a German whisky. Here, an evocation to the GI happens, in the absence of any phonetic and/or visual similarity between the disputed designation and the GI, by any partial incorporation of that indication in that designation, must take account of the conceptual proximity, if any, between them. In response to the second question of the preliminary ruling launched by the Regional Court, the Court concluded that for the purpose of establishing that there is an 'evocation' of a GI, account is not to be taken either of the context surrounding the disputed element, or, in particular, of the fact that that element is accompanied by an indication of the true origin of the product concerned.[57]

> 60 In view of the foregoing considerations, the answer to the second part of the second question is that Article 16(b) of Regulation No 110/2008 must be interpreted as meaning that, for the purpose of establishing that there is an 'evocation' of a registered geographical indication, account is not to be taken either of the context surrounding the disputed element, or, in particular, of the fact that that element is accompanied by an indication of the true origin of the product concerned.

As regards evocation, a paramount judgment is *Manchego*. It establishes overall conclusions and detailed lessons. The main one is that a registered name can be evoked by figurative signs. In the case, it is settled that the use in a label of a cheese a bony horse and landscapes with windmills and goats' tribes evokes to the *PGI Queso Manchego*. The GI's logo contains figurative elements consisting of a

[54] See Recital 35 Regulation (EU) 2024/1143.

[55] Guillem Carrau (2021), p. 391.

[56] Op cit. 16, para. 46; Op. cit. 44, para. 57 and 59; and Op. cit. 48, para. 48.

[57] Op. Cit 21, para. 60.

feature of the famous knight of the broken lance whose adventures have been related by Miguel de Cervantes in The Quixote. The recreation of the scene contained in the logo is composed by the Quixote riding his bony horse so-called Rocinante and among other elements, the windmills and the goats' tribes, which are also part of his stories. Therefore, the evocation is detected when any other cheese producer, including those established in the region but not accomplishing the GI rules, inserts any of the figurative elements and the term Rocinante in the label of their cheese and there is, therefore, a clear and direct link between the disputed elements.[58]

Some criticism has been expressed, as regards *Manchego* judgment, in the sense of accepting that non-compliance with (new) standards also means that any reference to geographic origin, even in an indirect, allusive form is forbidden? A strict prohibition of that kind would arguably risk resulting in a disproportional encroachment on the fundamental rights to impart and receive information, and to conduct one's business. As, according to the Court, using those rights as external balancing tools is not feasible, it is all the more important that the law itself provides for an explicit right to inform, in a manner conforming to honest business practices, about one's locality and the traditions observed. Further limitations drawing inspiration from trade mark law, including permissibility of comparative advertising, should be added; as a complement, an opening clause in the form of a general "due cause" defence could be envisaged[59]

A quotation is due as well to the *Cave de Tain* case law in which GC confirmed the absence of likelihood of confusion with the *GI PDO Cava*, because GI's protection does not extend against evocation when the differences between the signs, like figurative elements or conceptual meaning, precluded evocation. The visual and phonetic differences between the signs at issue were such as to preclude, in any event, the possibility of evocation.[60]

The EU Courts approach to the protection against evocation has been non-systematic and expansive although we need to mention that there are some cases in which the trend is not followed, like *Portsoy*.[61]

Evocation is a *sui generis* standard of protection within the EU's *sui generis* system for GI's protection. Absent in the international law, this concept first appeared at the EU level through Article 13.1. (b) of the now repealed Council Regulation (EEC) No 2081/92. In the first group of cases dealing with evocation, the European Courts characterized evocation with a reference to concepts pertaining to trade mark law—such as visual, phonetic and conceptual similarity—which are present in all trade mark-based analysis.[62]

A last word as regards the extension of the GIs' protection has been said in *Champanillo* case. As the Advocate General recalled in *Champanillo* case,[63] since

[58] Op. cit 51, para. 36, 40 and 41.

[59] Kur (2023), p. 92.

[60] T-774/16, Cave de Tain, ECLI:EU:T:2018:441, para. 44–52.

[61] Op cit. 25.

[62] Zafrilla Díaz-Marta and Kyrylenko (2021), p. 442.

[63] Op. cit 35, para. 58 and 59.

the finding of an evocation does not require proof of a likelihood of confusion, the relevant public, whose perception is important for the purpose of assessing whether the disputed sign can trigger an unlawful association with the registered name, does not consist solely of the group of persons for whom the products covered by that name are intended.

In the light of the objectives of the protection of fair competition and consumer protection pursued by the PDO and PGI protection mechanisms, the Court clarified that, in cases concerning the existence of an evocation of such designations, the national court is required to refer to the perception of an 'average consumer, who is reasonably well informed and reasonably observant and circumspect'[64] and that, in view of the need to ensure effective and uniform protection of those designations throughout the territory of the European Union, that concept covers European consumers[65] – and not only consumers of the Member State in which the product (or service) giving rise to the evocation of the registered name is manufactured (or provided).[66]

That being so, if the national court, which is responsible for conducting that assessment, relying on the presumed reaction of the consumer, concludes that when the consumer is confronted with the disputed sign, the 'image triggered directly in his mind' is that of the product covered by the registered name, the use of that sign is caught by the *ius prohibendi* of the producer group. However, if, in the opinion of that court, such an association of ideas is unlikely to occur, an evocation within the meaning of that provision must be ruled out. The concept of 'evocation' cannot, therefore, be calibrated.[67]

Finally, the Court concluded that 'evocation', first, does not require, as a preliminary condition, that the product protected by a PDO and the product or service covered by the disputed sign be identical or similar and, second, is established where the use of a name creates, in the mind of an average European consumer who is reasonably well informed and reasonably observant and circumspect, a sufficiently clear and direct link between that name and the PDO.

The existence of such a link can result from several aspects, in particular, the partial incorporation of the protected designation, the phonetic and visual relationship between the two names and the similarity resulting from it, and even in the absence of those aspects, from the conceptual similarity between the PDO and the name at issue or from a similarity between the products covered by that PDO and the products or services covered by that name.[68]

After the *Champanillo* case, it is time to reflect on GI protection in the field of services also from an active perspective. In fact, the recognition of such area of exclusivity may indeed have opened the door to further levels of exploitation of GIs

[64] See, by analogy, Op. cit. 48, para. 22 to 26.

[65] Op. cit. 48, para. 27; Op. cit. 51, para. 47 to 50; and Op. cit. 21, para. 59.

[66] Op cit 60, para. 27.

[67] AG Opinion Case C-783/19, Comité Interprofessionnel du Vin de Champagne v GB ECLI:EU:C:2021:350, para. 62 and 63.

[68] Op cit 30, para. 66.

in the distribution sector, through shops, bars, restaurants, agritourism and other activities offered to the public under the reserved trade name.20 From this point of view, the fact that the judgment in question concerned a service connected to the distribution of the PDO evoked does not reduce the scale of the issue, since this is indeed the sector of greatest commercial interest for typical geographical products.[69]

Turning to the determination of the applicable law, the Court held that acts of misuse, imitation or evocation of a PDO in relation to services also fall within the scope of the provision. It pointed out that the wording of Article 103(2)(b) expressly indicates that PDOs are protected against any misuse, imitation or evocation, 'even if the true origin of the product or service is indicated'. The Court further noted that Article 103(2)(b) provides for a wide-ranging protection which, according to recital 97 of the Wine Regulation, is designed to extend to all 'uses which take advantage of the reputation enjoyed by complying products' covered by the PDO. The Court also pointed to the second sentence of recital 97, which states that, to protect fair competition, the protection of PDOs 'should also extend to products and services not covered by the Regulation'. The Court concluded that Article 103(2)(b) prohibits *any* practice aimed at taking undue advantage of the reputation of a PDO through association with it. The Court's approach in this regard is in line with the Court's case law on the assessment of the risk of association under trade mark law.[70] However, the Court did not grasp the opportunity of the reference for a preliminary ruling in the *Champanillo* case to clarify whether the scope of protection of PDOs and PGIs against evocations under the Wine Regulation also extends to non-comparable goods. Neither did it look at whether the protection of PDOs/PGIs against 'uses', under Article 103(2)(a)(ii) of the Wine Regulation, covers uses in relation to services. From the reasoning followed by the Court, one can only gather that the answer to these two questions is positive.[71]

Regulation (EU) 2024/1143 draws attention to the concept of evocation. In recital 35, the legislator departs from the established case-law on the topic, by setting that '*evocation of a GI may arise, in particular, where a link with the product designated by the GI, including with reference to a term, sign, or other labelling or packaging device, is present in the mind of the average European consumer who is reasonably well-informed, observant and circumspect'*. Consequently, Article 26 Regulation (EU) 2024/1143 says that GI shall be protected, among other situations, against any misuse, imitation or evocation, even if the true origin of the products or services is indicated or if the protected name is translated, transcribed or transliterated or accompanied by an expression such as 'style', 'type', 'method', 'as produced in', 'imitation', 'flavour', 'like' or similar, including when those products are used as an ingredient.

It is very pertinent to highlight that the legislator has stated that GIs are protected against evocation, even if the true origin of the products or services is indicated. Furthermore, given the increased use of online intermediary services, the legislator

[69] Calabrese (2024), p. 348.

[70] C-252/07, Intel, EU:C:2008:655, para 42.

[71] Vrins (2021), p. 1171.

declared Article 26 Regulation (EU) 2024/1143 applicable to all the domain names accessible in the Union.

5.2.5 Any Other False or Misleading Indication as to the Provenance, Origin, Nature or Essential Qualities of the Product: The Impact of *Scotch Whisky Association*

In *Scotch Whisky Association*,[72] the Court had the opportunity of explaining when to establish a *'false or misleading indication as to the provenance, origin, nature or essential qualities ... of the product'*. This judgment is incredibly helpful to understand the current wording of Article 26(1)(c) Regulation (EU) 2024/1143.

As the Advocate General noted in point 92 of his Opinion in the referred case,[73] the expression 'false or misleading indication … on the description, presentation or labelling of the product' amounts to a list of various mediums on which the indication suspected of being false or misleading may be found. That does not permit the inference that that indication must be examined in conjunction with any other information that may be included on the description, presentation or labelling of the product concerned.

The Court took into consideration that, as has also been noted by the Advocate General in point 96 of his Opinion in *Scotch Whisky Association*, Article 16 of Regulation No 110/2008 contained a graduated list of prohibited conduct in which point (c) of Article 16 must be distinguished from points (a) and (b) thereof. Point (a) of Article 16 is limited to use of the protected geographical indication and point (b) to misuse, imitation or evocation. Point (c), however, widens the scope of the protection to include 'any other ... indication' (in other words, information provided to consumers that is included on the description, presentation or labelling of the product concerned) which, while not actually evoking the protected geographical indication, is 'false or misleading' as regards the links between the product concerned and that indication. In that regard, it must be stated, first, that the expression 'any other ... indication', used in Article 16(c) of Regulation No 110/2008, includes information that may be found in any form whatsoever on the description, presentation or labelling of the product concerned, inter alia in the form of words, an image or a container capable of providing information on the provenance, origin, nature or essential qualities of that product.

For the Court, it is sufficient that a false or misleading indication be included on one of the three mediums mentioned in that provision, namely 'on the description, presentation or labelling' of the product concerned, for it to be found that it is 'liable to convey a false impression as to [the product's] origin' within the meaning of that provision.

[72] Op cit. 21, para. 63 to 71.

[73] AG Opinion, C-44/17, ECLI:EU:C:2018:111, para. 92.

> 68 It follows from the foregoing that Article 16(c) of Regulation No 110/2008 affords extensive protection to registered geographical indications. If a false or misleading indication could nonetheless be permitted because it is accompanied by additional information relating, in particular, to the true origin of the product concerned, that provision would be deprived of practical effect.

Finally, the answer to the third question of the Regional Court is that Article 16(c) of Regulation No 110/2008 must be interpreted as meaning that the context in which the disputed element is used will not be taken into account for the purpose of establishing that there is a 'false or misleading indication'.

5.2.6 Any Other Practice Liable to Mislead the Consumer as to the True Origin of the Product: A Catch-All Clause Explained in *Morbier*

The expression any other practice liable to mislead the consumer as to the true origin of the product is a safety net clause for the GIs protection. As regards the conduct referred to in Article 26(1)(d) Regulation 2024/1143, in *Morbier*,[74] the Advocate General notes, in point 49 of his Opinion, that it is clear from the words 'any other practice' used in those provisions that they are intended to cover any conduct not already covered by the other provisions of those articles and thus to tighten the system of protection for registered names.

This provision can be defined as a system shutdown provision, which covers any practices not explicitly mentioned in the other points of Article 26 of Regulation (EU) 2024/1143, with the primary objective of preventing consumer deception regarding the true origin of the product.[75]

Consequently, the scope of the provision is not limited to prohibiting the use of the registered name itself but have a wider scope. It does not prohibit solely the use by a third party of a registered name. For the Court, it must be interpreted as prohibiting the presentation of a product protected by a designation of origin, in particular the reproduction of the shape or the appearance which are characteristic of it, which is liable to mislead the consumer as to the true origin of the product, even if the registered name is not used. It will be necessary to assess whether such reproduction may mislead the European consumer, who is reasonably well informed and reasonably observant and circumspect, considering all the relevant factors in the case. In particular, as regards, an element of the appearance of the product covered by the registered name, it is necessary, in particular, to assess whether that element constitutes a baseline characteristic which is particularly distinctive of that product so that its reproduction may, in conjunction with all the relevant factors in the case in point,

[74] Op. cit 22.

[75] Arroyo Aparicio (2022), p. 494.

lead the consumer to believe that the product containing that reproduction is a product covered by that registered name.[76]

In *Morbier*, the preliminary ruling was focused on the possibility of misleading the consumer by the existence of the one black line in the middle of the cheese visually identical to the one which characterises the *Morbier* cheese. Technically the Court of Cassation decided to stay proceedings and refer the question to the Court to know whether the EU legislator prohibits solely the use by a third party of the registered name, or that the EU legislator prohibits the presentation of a product protected by a designation of origin, in particular the reproduction of the shape or the appearance which are characteristic of it, which is liable to mislead the consumer as to the true origin of the product, even if the registered name is not used in the presentation of the other product.

The Court interpreted that the extension of the GIs protection goes until the extreme of prohibiting the reproduction of the shape or appearance characterising a product covered by a registered name where that reproduction was liable to lead the consumer to believe that the product in question is covered by that registered name. For the Court, it was necessary to assess whether such reproduction may mislead the European consumer, who is reasonably well informed and reasonably observant and circumspect, considering all the relevant factors in the case.

This has been criticised because while in trade mark law protection of product appearances is heavily qualified so as to prevent monopolisation of elements that should remain free to be used by competitors, such aspects were not given any consideration in the EU Court's assessment in *Morbier* of whether GI protection can extend to elements of appearance.[77] In its ground-breaking decision, the Court of Justice of the European Union has extended protection of a GI to the reproduction of shape or appearance of the product itself. The interesting bit is the CJEU's extension of scope of Article 13(1)(d) to the shape or appearance characteristic of a product without need for making reference to the protected name. The Court had yet again taken the scope of protection for GIs to a new level—and the decision could be seen as the peak of a string of decisions, which included *Scotch Whisky Association* and *Queso Manchego*. However, whereas in the latter decisions the infringing acts related to the GIs as such, this was not the case here. The CJEU had thus opened a Pandora's box with a need to clarify where to draw the line. Until then, product clearance will be extremely difficult as it is not enough to look at protected names.[78]

5.2.7 An Extension of the Protection Under Unfair Competition Law: A Reflection on the Issue from *Salame Felino*

The *ex lege* protection of the GI, under the EU quality food scheme is complementary to the protection that the GI has, as any other intellectual property right, under

[76] Op. cit 22, para. 41.

[77] Kur (2023), p. 91.

[78] Günzel (2021), p. 294.

the rules of the Unfair Competition Law. Article 22(2) TRIPS Agreement offers an example of this conception of an extended umbrella of protection for the geographical indications.

The use of geographical indications or of any other means in the designation or presentation of a product which indicate or suggest that such product originates in a locality other than its true place of origin, or that the product has such qualities as are typical of products originating in a locality designated by a GI, could constitute an act of unfair competition where it is liable to mislead the public.

This question was raised to the Court in *Salame Felino*[79] concerning the conditions under which a geographical designation may be used to designate a product manufactured outside of the relevant zone where that designation is not recognised as a protected designation of origin or as a protected geographical indication within the meaning of EU Law. At stage, the Court said that it is for the national court to satisfy itself, first, that the application of the national legislation in question does not have the purpose or effect of guaranteeing to consumers that products marketed under a specific geographic designation have a particular quality or characteristic and, secondly, that the implementation of that legislation pursues an overriding requirement in the public interest, is carried out in a non-discriminatory manner, is suitable for securing the attainment of the objective pursued and does not go beyond what is necessary in order to attain that objective. It confirmed that it does not afford protection to a geographical designation which has not obtained a EU registration, but that that geographical designation may be protected, should the case arise, under national legislation concerning geographical designations relating to products for which there is no specific link between their characteristics and their geographical origin, provided, however, first, that the implementation of that legislation does not undermine the objectives pursued by Regulation No 2081/92 and, secondly, that it does not contravene the principle of the free movement of goods under Article 28 EC, matters which fall to be determined by the national court.[80]

In *Scotch Whisky Association*, the Court stated that it must be reminded that to prevent the misuse of protected geographical indications, is not only in the interests of consumers, but also in the interests of producers who have striven to guarantee the qualities expected of products lawfully bearing such indications. In that context, it is not allowed, more specifically, operators from making commercial use of a registered geographical indication in respect of products that are not covered by the registration, with the aim of taking unfair advantage of that GI.[81]

> 69 Finally, as has already been stated in paragraph 38 of this judgment, the purpose of Regulation No 110/2008 and, in particular, of Article 16 thereof, is to protect registered geographical indications, both in the interests of consumers who should not be misled by inappropriate indications, and in the interests of economic operators which bear higher costs in order to guarantee the quality of products that lawfully bear protected geographical indications. Those operators must be protected against acts of unfair competition.

---

[79] C-35/13, Assica and Krafts Foods Italia, ECLI:EU:C:2014:306, para. 7, 38 and 39.

[80] Guillem Carrau (2022), p. 565.

[81] Op cit 21, para. 38.

In this context, the goal of both disciplines, trade mark law and unfair competition law is different. Trade mark law protects directly the interest of producers, however unfair competition law is more for protecting consumers from the unacceptable use of a geographical indication.[82]

In *Champanillo*, one of the topics of discussion was whether the protection against evocation provided for in EU Law requires a finding of unfair competition. As the Advocate General recalled in his Opinion,[83] it is possible that the same conduct may simultaneously satisfy the conditions of a practice prohibited under Article 103(2)(b) of Regulation No 1308/2013 and an act of unfair competition within the meaning of the applicable national law, the scope of that provision is broader and is not limited to cases where that conduct is engaged in by a competitor. However, the protection is not limited to cases where the practice giving rise to evocation satisfies the conditions for an act of unfair competition within the meaning of the relevant provisions of the applicable national law. The Court concluded that EU Law states a specific protection which applies independently to the national unfair competition rules.[84]

5.3 Conclusions

The examination of the GC/CJ pertinent judgments like *Feta, Grana Padano, Edam Holland,* and *Aceto Balsamico di Modena* permits drawing conclusions on the complexities of GI protection, revealing a nuanced framework that balances origin and specificity with the threat of genericity. Through key cases, such *Bavaria, Chianti,* and *Scotch Whisky Association,* the Court has established clear guidelines for GI protection, including the interpretation of the prohibition of direct and indirect use, misuse and misleading practices. The *BNIC* criteria establishes the way of assessing comparability of products, which is a key finding in the context of the delimitation of the extension of the GIs protection. As regards evocation, *Manchego, Morbier* and *Champanillo* are the landmark cases that make technical findings on the span of GIs protection. On the top of that, the extension of the GIs protection under unfair competition law is discussed in *Salame Felino*.

References

Arroyo Aparicio A (2022) Denominaciones de origen: reproducción de la forma o de la apariencia del producto como infracción (Designations of origin: reproduction of the form or the appereance of the producto as an infraction) in Competencia, propiedad intelectual y tutela de consumidores en el sector agroalimentario / coord. por Vanessa Jiménez Serranía; Fernando Carbajo Cascón (dir.), Tirant lo Blanch, pp 475–495

[82] O'Connor (2004), p. 203.

[83] Op. cit 35, para. 72–76.

[84] Op cit 30, para. 70.

Calabrese B (2024) The evolving protection of geographical indications against services: "brand" new world? IIC 55:348–367

De Almeida AR, Carls S (2021) The criteria to qualify a geographical term as generic: are we moving from a European to a US perspective? IIC 52:444–467

Guillem Carrau J (2021) Denominaciones geográficas de calidad. Estudio de su régimen jurídico en la OMC, la Unión europea y el Derecho español (Quality Geographical Indications. Study of the legal framework in the WTO, the EU and Spain, Tirant lo blanch, 2 ed, Valencia

Guillem Carrau J (2022) How much protection for Geographical Indications (GIs)? A jigsaw of case law is falling into place in the EU. EIPR 44(9):556–566

Günzel O (2021) Ashes to ashes for cheese featuring the 'Morbier' visual appearance? J Intellect Property Law Pract 16(4–5):293–294

Kur A (2023) No strings attached to GIs? About a blind spot in the (academic) discourse on limitations and fundamental rights. IIC 54:87–94

Mantrov (2012) Protection norms of indications of geographical origin in the applicable EU regulations – recent changes and the necessity for further unification. IIC 43(2):174–202

O'Connor B (2004) The law of geographical indications. Cameron May, London

Vrins O (2021) No champagne for 'Champanillo': the protection of PDOs and PGIs against evocation for services. J Intellect Property Law Pract 16(11):1171–1174

Zafrilla Díaz-Marta V, Kyrylenko A (2021) The ever-growing scope of Geographical Indications' evocation: from Gorgonzola to Morbier. J Intellect Property Law Pract 16(4–5):442–449

6 Labelling Rules for Geographical Indications of Provenance

Abstract

GIs labelling rules are an expression of the exclusive use of the GI and they are interpreted in connection to labelling rules. In this chapter, the compulsory rules on the use of Union symbols on the packaging (labelling and advertising material) are examined within the context of the Commission Guidelines on labelling practices. *Champagner sorbet* and *Weingut* cases serve as example of the relevance of the enforcement of the GIs as any other intellectual property right and to understand how it works GI when is used as an ingredient and to illustrate about some specificities of labelling and presentation in GI wines, respectively.

The system of GIs aims at enabling consumers to make more informed purchasing choices and, through labelling and advertising, helping them to correctly identify their products on the market. Direct or indirect references to the geographical origin of a product, i.e., its place of production, manufacture, processing, harvesting, or extraction, are common in commercial transactions.

To paraphrase Mattera,[1] in commerce, such references appear either directly, through the designation of the product or the product itself with a geographical name, a localizing adjective, or a mention of the place of manufacture, or indirectly, by evoking a specific geographical origin, as occurs with a denomination in a specific language that evokes the country where it is spoken. This indirect indication does not necessarily imply specific characteristics or peculiar quality. Unlike direct references, indirect references were not initially considered eligible for protection under EU law, but they were in some EU states, such as Germany, which led to the

[1] Mattera (1991), p. 210.

© The Author(s), under exclusive license to Springer Nature Switzerland AG 2026

J. Guillem Carrau, *Landmark EU Geographical Indications Case Law*, Law for Professionals, https://doi.org/10.1007/978-3-032-26101-4_6

inclusion of some of them in the 2012 reform of the European agri-food quality system.[2]

Although the motivation of the EU GIs system is not to grant an adequate labelling neither accomplishing phytosanitary rules nor to have a legal protection similar to a trade mark, the EU GIs scheme takes into account the consumer's interests by providing them a better information about production methods and origin of the product.

Likewise, the visibility of the EU GI system is linked to labelling and the compulsory use of the Union symbols refrains this statement. Besides, the labelling is also the way the legislator highlights the exclusive right to use the GI in the commerce. The use of the protected term is undoubtedly an expression of the exclusive character of the intellectual property right that manifests in the presentation of the product (labelling and advertisement) and the documentation and the use of the Union symbols. In fact, the EU legislator does not expressly declare the exclusive right of use of the GI in the markets. This can be deduced, and it will be concretised in the use of the term in the commerce when presenting the product (labelling and advertising) and with the documentation and the use of the characteristic logos.[3]

In any case, the protection extends to labelling as it includes all the steps of commercialisation, advertisement, labelling and any other type or commercial document.[4]

6.1 The Connection with the Food Labelling Policy

The labelling of spirit drinks and agricultural products is subject to the general rules laid down in Regulation (EU) No 1169/2011[5] (also known as FIC Regulation (Food Information to Consumers))[6] which establishes general principles, requirements, and responsibilities governing food information, in particular labelling. The FIC Regulation applies to food business operators at all stages of the food chain and applies to all foods intended for the final consumer. In particular, as stated in Article 1(4) of the FIC Regulation, it applies without prejudice to labelling requirements in specific EU provisions applicable to particular foods, including GI and

[2] Guillem Carrau (2021), p. 38.

[3] Maroño Gargallo (2002), p. 262.

[4] Guillem Carrau (2021), p. 634.

[5] Article 2 (1) (c) Regulation (UE) 2024/1143 states that labelling' means, in respect of all products falling within the scope of this Regulation, labelling as defined in Article 2(2), point (j), of Regulation (EU) No 1169/2011.

[6] Regulation (EU) No 1169/2011 of the European Parliament and of the Council of 25 October 2011 on the provision of food information to consumers, amending Regulations (EC) No 1924/2006 and (EC) No 1925/2006 of the European Parliament and of the Council, and repealing Commission Directive 87/250/EEC, Council Directive 90/496/EEC, Commission Directive 1999/10/EC, Directive 2000/13/EC of the European Parliament and of the Council, Commission Directives 2002/67/EC and 2008/5/EC and Commission Regulation (EC) No 608/2004, OJ L 304, 22.11.2011.

TSG products. The inclusion of quality products in the scope of the FIC Regulation ensures coherence between the two policy areas.[7]

Both legal schemes aim at preventing labelling that may confuse or mislead consumers. In both cases, the legislation lays down labelling rules to ensure that consumers are provided with reliable information that enables them to make informed choices and avoid being misled. Although EU quality policy relating to GIs is not per se an origin labelling system, it links food and beverages to a specific geographical area, to which the name of the product refers. Therefore, the link with territory and the indication of origin represents the key feature of the quality schemes.[8] Having said that, the policies tackle consumer information from slightly different angles: while the FIC Regulation provides a first layer of requirements applicable to all food products, GI policy concerns the value-adding attributes of GI products linked to their origin. Moreover, unlike the quality policy, the FIC Regulation makes a link between the information provided and the protection of consumers' health, thus covering nutrition aspects as well. However, various other legal frameworks refer to the indication of the origin of food products.[9]

In the field of GIs, specific labelling rules may be imposed by the applicant. If this is the case, the product specification and the single document must also provide specific information on them. Where the single document includes specific requirements on packaging and labelling or other applicable requirements, provided for by the product specification, any restriction resulting from those requirements shall be accompanied by a summary of the product-specific justification contained in the correspondent product specification.[10]

Regarding GI wines, there are optional particulars, like the vintage year or the name of the grape variety, and compulsory mentions as the producer, the bottler or the vendor. Wine compulsory particulars referred to in Article 119 of Regulation (EU) No 1308/2013 must appear in the same field of vision on the container, in such

[7] GIs fall within the scope of Article 26(3) of the FIC Regulation. In view of the fact that for these product designations an intrinsic link between product characteristics and geographical origin exists, they are regulated by specific rules, including rules on labelling, and taking into consideration the specific character of these names as intellectual property rights, it is necessary to further examine how the origin of the primary ingredient provided by Article 26(3) of the FIC Regulation should be indicated for said names. For this reason, Article 1(2) of Implementing Regulation (EU) 2018/775 specifies that GI products do not fall under its scope. Recital 6 clarifies that, while the general requirements of Article 26(3) of the FIC Regulation apply to all food products, the relevant implementing rules, related to the primary ingredient, require further examination and will be adopted at a later stage.

[8] See Commission Staff working document Evaluation of GIs and TSG in the EU, SWD/2021/0427 final, accessible in https://eur-lex.europa.eu/legal-content/EN/TXT/?uri=SWD:2021:427:FIN.

[9] Article 26 of the FIC Regulation establishes general requirements with regard to the indication of products' country of origin or place of provenance. These must be indicated where failure to do so this might mislead the consumer. Beyond this general requirement, EU law also requires an indication of origin for certain specific foodstuffs, such as fresh fruit and vegetables, fishery products, honey, olive oil, wine and eggs and for beef and beef products, pig meat, poultry and sheep and goat meat.

[10] Article 4(2) Commission Implementing Regulation (EU) 2025/26.

a way as to be simultaneously legible without having to turn the container, in indelible characters and must be clearly distinguishable from surrounding text or graphics.[11] Some specific rules apply when there is an indication of provenance as 'wine of…' or 'European Union wine'.[12]

6.2 The Use of Union Symbols in GIs Products

There is limited public awareness of Union symbols (e.g. only 20% of consumers surveyed recognised the PGI logo), and there is confusion between the different labelling schemes.[13] Despite of, the use of the Union symbol is not mandatory for wine and spirits, it is for the applicant to decide if its use is mandatory or not.[14] In the case of foodstuffs, the us is mandatory according to article 37(3) Regulation (EU) 2024/1143.

The use of Union symbols on the packaging (labelling and advertising material) of agricultural products designated by a GI aims to make that category of products, and the guarantees attached to them, better known to consumers and to permit easier identification of those products on the market, thereby facilitating checks with some specifications for spirit drinks and third country GIs.[15] Same rules apply for craft GIs.[16]

The Union symbols referred to in Article 37(2) and Article 70(1) of Regulation (EU) 2024/1143 must be reproduced as laid down in Annex XVII of Commission Implementing Regulation (EU) 2025/26.[17] Where the Union symbols, indications or corresponding abbreviations as referred to in Articles 37 and 70 of Regulation (EU) 2024/1143 appear on the labelling of a product, they must be accompanied by the

[11] See Article 40 Regulation (EU) 2019/33.

[12] See Article 45 Regulation (EU) 2019/33.

[13] Commission Staff working document Evaluation of GIs and TSG in the EU, SWD/2021/0427 final, accessible in https://eur-lex.europa.eu/legal-content/EN/TXT/?uri=SWD:2021:427:FIN.

[14] Commission Delegated Regulation (EU) 2025/27 of 30 October 2024 supplementing Regulation (EU) 2024/1143 of the European Parliament and of the Council with rules concerning the registration and the protection of geographical indications, traditional specialities guaranteed and optional quality terms and repealing Delegated Regulation (EU) No 664/2014, OJ L, 2025/27, 15.1.2025.

[15] In view of their specific nature, special provisions concerning labelling are maintained for spirit drinks after 2024 reform. The use of Union symbols or indications should remain voluntary for third country geographical indications and designations of origin for agricultural products. See Recital 49 Regulation (EU) 2024/1143.

[16] Guillem Carrau (2023), p. 725.

[17] Commission Implementing Regulation (EU) 2025/26 of 30 October 2024 laying down rules for the application of Regulation (EU) 2024/1143 of the European Parliament and of the Council as regards registrations, amendments, cancellations, enforcement of the protection, labelling and communication in respect of geographical indications and traditional specialities guaranteed, and amending Implementing Regulation (EU) 2019/34 as regards geographical indications in the wine sector, and repealing Implementing Regulations (EU) No 668/2014 and (EU) 2021/1236, OJ L, 2025/26, 15.1.2025.

registered name.[18] Moreover, the indications 'PROTECTED DESIGNATION OF ORIGIN', 'PROTECTED GEOGRAPHICAL INDICATION' and 'TRADITIONAL SPECIALITY GUARANTEED' within the symbol can be used in any of the official languages of the Union as laid down in Annex XVII of Commission Implementing Regulation (EU) 2025/26.

According to Regulation (EU) 2024/1143, depictions of the geographical area of origin referred to in the product specification and text, graphics or symbols referring to the Member State and the region in which that geographical area of origin is located can appear in the labelling as well. However, Union symbols indicating the PDO or the PGI, the Union indications 'protected designation of origin' or 'protected geographical indication' and the Union abbreviations 'PDO' or 'PGI' may appear on the labelling only after the publication of the decision conferring protection on that designation of origin or geographical indication.[19]

To sum up, there are a set of mentions that can appear in the labelling of a GI product: the Union symbol, the indication protected geographical indication, the abbreviation PGI, a specific logo for the GI concerned (if included in the single document/product specification as obligatory labelling), depictions of the demarcated geographical area and text, graphics, or symbols referring to the Member State or the region where the area of origin is located. Besides, to give visibility to the producers of products designated by GIs, it should be mandatory to indicate on the label the name of the producer or, in the case of agricultural products, the name of the operator.

6.3 Commission Guidelines on the Labelling of Foodstuffs Using GIs

Guidelines play an important role in contemporary public administration. Rawlings considers that, since Megarry first identified "quasi-legislation" in the 1940s, the use of guidelines has only expanded. Guidelines are now pervasive, found at every level of public administration.[20] As regards the guidelines as instrument of soft law, despite the propagation of guidelines in recent decades, their legal effects are often difficult to classify.[21]

Though praised for their flexibility, adaptability and 'rapid reaction' capacity, Commission soft law instruments are controversial, as they might arguably entail far-reaching consequences due to their contentious 'hardening,' i.e. the capacity to produce (not negligible) legal and practical effects, despite their legally non-binding nature enshrined in Article 288 TFUE. There are four argumentative parameters that

[18] Article 34(2) Commission Implementing Regulation (EU) 2025/26.

[19] See Article 22 Regulation (EU) 2019/33.

[20] Rawlings (2015) and Megarry (1944).

[21] Daly (2023), p. 211.

require extensive attention for a better drafting and assessment of the Commission soft law instruments: *content*, *design*, *effectiveness* and *soundness*.[22]

In 2010, the European Commission published a Communication[23] on the labelling of foodstuffs using PDOs and PGIs with the aim of illustrating the legislative provisions applicable in this area and '*to help economic operators define their room for manoeuvre*'. In fact, the Commission draw the voluntary Guidelines to give some light on the following points: the conditions under which names registered as a PDO or PGI can be used in the labelling, presentation and advertising of foodstuffs containing such names as ingredients; and good practice to ensure that names registered as a PDO or PGI and employed as ingredients in food products are not used in a manner that damages the reputation of the product benefiting from such a designation or misleads consumers as to the composition of the product produced.

These Guidelines contain a set of recommendations amongst other, firstly that a name registered as a PDO or PGI may legitimately be included in the list of ingredients of a foodstuff; secondly that a name registered as a PDO or PGI may be mentioned in or close to the trade name of a foodstuff incorporating products benefiting from a registered name, as well as in the labelling, presentation and advertising relating to that foodstuff under certain conditions;[24]; thirdly, that the European Union terms, abbreviations or symbols accompanying the registered name should be used in labelling, within or close to the trade name or in the list of ingredients of the foodstuff only if it is made clear that the said foodstuff is not itself a PDO or PGI; and fourthly that if an ingredient comparable to an ingredient benefiting from a PDO/PGI has been incorporated in a foodstuff, the name registered as a PDO/PGI should appear only in the list of ingredients, in accordance with rules similar to those applicable to the other ingredients mentioned.

All in all, the guidelines were not deemed to constitute a legally binding interpretation of EU legislation on PDOs and PGIs or the Labelling Directive. Indeed, for the Commission, such an interpretation falls solely within the remit of the European

---

[22] Andone and Coman-Kund (2022), p. 22.

[23] Commission Communication — Guidelines on the labelling of foodstuffs using protected designations of origin (PDOs) or protected geographical indications (PGIs) as ingredients, OJ C 341, 16/12/2010.

[24] The foodstuff in question should not contain any other 'comparable ingredient', i.e. any other ingredient which may partially or totally replace the ingredient benefiting from a PDO or PGI. As a non-restrictive example of the concept of 'comparable ingredient', the Commission considers that a blue-veined cheese (commonly known as 'blue cheese') could be considered comparable to 'Roquefort' cheese. This ingredient should also be used in sufficient quantities to confer an essential characteristic on the foodstuff concerned. However, given the wide range of possible scenarios, the Commission is not able to suggest a minimum percentage to be uniformly applied. As an example, the incorporation of a minimum amount of a spice benefiting from a PDO/PGI in a foodstuff could, if appropriate, be sufficient to confer an essential characteristic on that foodstuff. By contrast, the incorporation of a minimum amount of meat benefiting from a PDO/PGI in a foodstuff would not a priori be sufficient to confer an essential characteristic on a foodstuff. Finally, the percentage of incorporation of an ingredient with a PDO or PGI should ideally be indicated in or in close proximity to the trade name of the relevant foodstuff or, failing that, in the list of ingredients, in direct relation to the ingredient in question.

Court of Justice; furthermore, the issue of whether a specific product's labelling could mislead purchasers or consumers, or any decision regarding the potentially misleading nature of a trade name is the responsibility of domestic courts.[25]

According to Article 95(1) (k) Regulation (UE) 2024/1143, the single document of a GI must contain the specific rules concerning packaging and labelling as well as any other essential relevant requirements. The single document shall contain the product specification, namely the name to be registered as a designation of origin or geographical indication, a description of the product, including, where appropriate, specific rules concerning packaging and labelling and a concise definition of the geographical area; and a description of the link between the product and the geographical environment or geographical origin.[26]

6.4 *Champagner sorbet*: Exploiting the Reputation of GIs as Ingredients

In *Champagner sorbet*, the Court of Justice had the opportunity to clarify how to apply the labelling rules when using GIs as ingredients. The preliminary ruling was about a German supermarket chain selling a sorbet containing Champagne, which was distributed under the name 'Champagner Sorbet'. In fact, what was discussed was whether that conduct was lawful or whether the manufacturer and the distributor of the sorbet were taking advantage of the reputation of the French sparkling wine which benefits from a PDO.

In this context, the scope of the protection given by the EU legislator to a GIs covers a situation where a PDO, such as 'Champagne', is used as part of the name under which a foodstuff is sold, such as 'Champagner Sorbet', and where that foodstuff does not correspond to the product specifications for that PDO but contains an ingredient which does correspond to those specifications.

The Court concluded, firstly, that the use of a PDO as part of the name under which is sold a foodstuff that does not correspond to the product specifications for that PDO but contains an ingredient that does correspond to those specifications, such as 'Champagner Sorbet', constitutes exploitation of the reputation of a PDO, within the meaning of those provisions, if that foodstuff does not have, as one of its essential characteristics, a taste attributable primarily to the presence of that ingredient in the composition of the foodstuff; secondly that the use of a PDO as part of the name under which is sold a foodstuff that does not correspond to the product specifications for that PDO but contains an ingredient that does correspond to those specifications, such as 'Champagner Sorbet', does not constitute misuse, imitation or evocation within the meaning of the EU legal framework for GIs.[27]

While the CJEU most often adopts a pointedly pro-protection attitude towards GIs, there are also exceptions to that scheme. Most notably this concerns the

[25] C-446/07 Alberto Severi v Regione Emilia Romagna (2009) ECR I-8041, para. 60.

[26] Article 50 Regulation (UE) 2024/1143.

[27] C-393/16 - Comité Interprofessionnel du Vin de Champagne, ECLI:EU:C:2017:991, pp. 62–63.

decision that using the (German) name "Champagner Sorbet" for ice cream containing champagne does not amount to misuse, imitation or evocation of the PDO, and that the renown enjoyed by champagne is only exploited in an unfair manner if the taste of the product is not essentially attributable to the presence of that ingredient. The decision thus provided an opening for the use of GIs in a descriptive form. The pending proposals address the issue by stipulating that reference to GIs as ingredients of a processed product is permitted, provided that the use is made in accordance with honest commercial practices and does not weaken, dilute or detrimentally affect the reputation of the geographical indication.[28]

Champagner Sorbet case inspired the EU legislator when approving Article 27 Regulation (EU) 2024/1143. GIs labelling is subject to the general rules at EU Level.[29] As indicated by the EU legislator, Regulation (EU) No 1169/2011[30] contains these general rules and the provisions aimed at preventing labelling that may confuse or mislead consumers. Recital 36 of Regulation (EU) 2024/1141 is self-explanatory of the intentions of the legislator when refers to the labelling of the products with GI ingredient: fair commercial practices, not dilution of GIs reputation and involvement of the recognised producer group when the GI is used in pre-package food.

> … in light of commercial practices and Union case-law, clarity is required on the use of a geographical indication in the sale name of a processed product of which the product designated by the geographical indication is an ingredient. It should be ensured that such use is made in accordance with fair commercial practices and does not weaken, dilute or is not detrimental to the reputation of the product bearing the geographical indication. To that end, conditions for the qualities attributed by the geographical indication as ingredient to the processed food should be added. Moreover, the producers of prepacked food should notify the recognised producer group, where such a group exists, before starting to use the geographical indication in the name of the prepacked food. Such approach is in line with the objectives of strengthening the protection of the geographical indications and enhancing the role of the recognised producer groups.

In this context, the content of the COM Guidelines is now reflected in the new provision regulating 'ingredients': article 27 Regulation (EU) 2024/1143 provides that the GI designating a product used as an ingredient in a processed product may be used in the name of that processed product, or in its labelling, or in its advertising material where:

---

[28] Kur (2023), p. 92.

[29] According to Article 2.1 (c) Regulation (EU) 2024/1141 labelling' means, in respect of all products falling within the scope of this Regulation, labelling as defined in Article 2(2), point (j), of Regulation (EU) No 1169/201.

[30] Regulation (EU) No 1169/2011 of the European Parliament and of the Council of 25 October 2011 on the provision of food information to consumers, amending Regulations (EC) No 1924/2006 and (EC) No 1925/2006 of the European Parliament and of the Council, and repealing Commission Directive 87/250/EEC, Council Directive 90/496/EEC, Commission Directive 1999/10/EC, Directive 2000/13/EC of the European Parliament and of the Council, Commission Directives 2002/67/EC and 2008/5/EC and Commission Regulation (EC) No 608/2004 (OJ L 304, 22.11.2011, p. 18).

(a) the processed product does not contain any other product comparable to the ingredient designated by the GI;
(b) ingredient designated by the GI is used in sufficient quantities to confer an essential characteristic on the processed product concerned; and
(c) the percentage of the ingredient designated by the GI in the processed product is indicated in the label.[31]

In addition, producers of a prepacked food, as defined in Article 2(2), point (e), of Regulation (EU) No 1169/2011, containing as an ingredient a product designated by a GI, who want to use that GI in the name of that prepacked food, including in advertising material, shall give a prior written notification to the recognised producer group where such a group exists for that geographical indication. Those producers shall include in that notification the information that demonstrates that the conditions listed in the precedent paragraph are complied with and shall act accordingly. In reply, the recognised producer group shall acknowledge receipt of that notification in writing within four months. The producer of prepacked food may start using the GI in the name of the prepacked food following the receipt of that acknowledgment or after the expiry of that time period, whichever occurs first. The recognised producer group may attach to that acknowledgement non-binding information on the use of the GI concerned and the recognised producer group and the producer of prepacked food can conclude a contractual agreement about the specific technical and visual aspects of how the geographical indication of the ingredient is presented in the name of the prepacked food in labelling, elsewhere than in the list of ingredients, or in advertising material. Indications and abbreviations may be used in the labelling and advertising material of processed products when the geographical indication refers to an ingredient thereof. In that case, the indication or abbreviation shall be placed next to the name of the ingredient that is clearly identified as an ingredient.

A good practice to be implemented by the GIs recognised producer group is to describe in the single document any relevant rule related to the labelling of the protected products or their use as an ingredient.

Last, the use in the labelling of the GI product of Union symbols indicating the PDO or PGI and the indications 'protected designation of origin', 'protected geographical indication' and 'geographical indication' and the abbreviations 'PDO' or 'PGI' as relevant is regulated in Article 37 Regulation (EU) 2024/1143, but only after the publication of the act of registration of that GI.

[31] This article does not apply to spirit drinks.

6.5 *Weingut*: Labelling and Presentation in the Wine Sector

EU law regulates in detail PDO, PGIs and traditional terms in the wine sector, as well as the labelling and presentation of wines.[32] O'Connors describes how the Court of Justice has regularly interpreted the EU provisions laying down denomination rules for wines,[33] for example the use of terms as 'crémant',[34] and 'chateau',[35] the reference to the method of production known as 'methode champanoise'[36] or affixing decoration unconnected with the wine marketed.[37] In this last case, Michèle Voisine was charged before the Tribunal de Police in Bordeaux on the ground that bottles of Bordeaux wine offered for sale carried labelling which included information not permitted and the presentation of a number of bottles of Champagne was contrary to law. Since names of towns appeared on the bottles which could be confused with the designation of origin of those geographical localities or the 'Romorantin' vine variety, their presentation was allegedly liable to mislead purchasers as to the provenance of the wine or the vine variety. The Court concluded that the definition of labelling covers decoration or advertising which is unconnected with the wine concerned.[38]

At this stage, *Weingut* case[39] helps to illustrate about some specificities of the EU GIs scheme and wine labelling. In the case at hand, the preliminary ruling was about a situation in which the owner of a wine-growing holding located in Zell (Germany), in the Moselle region, produced wine from grapes coming not only from its own vineyards but also from other vineyards which it leases and were labelled with the terms 'Weingut' (wine-growing estate) and 'Gutsabfüllung' (bottled at the estate).

The Federal administration (Land Rheinland-Pfalz) took the view that, in those circumstances, the winemaker could not use the indications 'Weingut' and 'Gutsabfüllung' for wine made on the premises of the holding of other wine-grower, given the lack of autonomy of the permanent establishment and the fact that the applicant in the main proceedings did not employ its own staff for the purposes of the pressing operation. The decision was appealed, and the first instance upheld the action, primarily on the ground that responsibility for the actual management,

[32] See for instance, Commission Notice – Questions and answers on the implementation of new EU wine labelling provisions following the amendment of Regulation (EU) No 1308/2013 of the European Parliament and of the Council and Commission Delegated Regulation (EU) 2019/33, C/2023/7808, OJ C, C/2023/1190, 24.11.2023. This document provides technical replies to questions that the Commission services have received, and have been discussed with experts from Member States, in relation to the application of the rules on labelling of wines introduced by Regulation (EU) 2021/2117 of the European Parliament and of the Council (1) amending Regulation (EU) No 1308/2013.

[33] O'Connors (2004), p. 161.

[34] C-309/89 - Codorniu v Council ECLI:EU:C:1994:197.

[35] C-403/92 - Baux v Château de Calce ECLI:EU:C:1994:269.

[36] C-306/93 - SMW Winzersekt v Land Rheinland-Pfalz ECLI:EU:C:1994:407.

[37] C-46/94 - Voisine ECLI:EU:C:1995:221.

[38] Op. Cit. 28, para. 30.

[39] C-354/22, Weingut A, ECLI:EU:C:2023:916.

continuous supervision and exclusive responsibility for winemaking lay with the applicant in the main proceedings. The Judgment was confirmed in second instance and, after a revision appeal, the Federal administrative Court was requested to solve an appeal on a point of law. In a preliminary ruling request, the Federal administrative Court submitted several questions to the Court about the interpretation of Article 54(1) of Delegated Regulation 2019/33 to a situation in which the winemaking was entirely carried out on the eponymous wine-growing holding.

The General Court concluded that a winemaker may indicate his wine-growing estate in the labelling of his wine, even if the process of grape cultivation and grape pressing took place on the wine-growing estate of another winemaker.[40] In addition, the General Court clarified the definition of winemaking in the referred legal context. For the Court, it must be accepted that winemaking is entirely carried out on the eponymous wine-growing holding, for the purposes of that provision, even if the pressing operation was performed by staff of the wine-growing holding which rented out the pressing facility to the eponymous wine-growing holding, provided that the owner of the latter holding assumes actual management, close and continuous supervision and responsibility for that operation. The fact that the wine-growing holding which rented out the winepress facility has an interest of its own in the manner in which the pressing is performed, in particular by reason of a contractual clause relating to a price supplement linked to yield and quality per hectolitre of wine, has no bearing on whether winemaking may be regarded as having been carried out on the eponymous wine-growing holding.

6.6 Conclusions

Based on a case law analysis, this chapter was found useful to explain that the labelling rules underpin the exclusive use of GIs including the compulsory use of Union symbols. Nevertheless, the analysis must be done equally as regards the Food Labelling Regulation.

Through cases such as *Champagner sorbet* and *Weingut*, it has been highlighted the importance of enforcing GI rights and the specific considerations that apply when GIs are used as ingredients or in wine labelling, demonstrating the practical implications of GI protection in the marketplace.

References

Andone C, Coman-Kund F (2022) Persuasive rather than 'binding' EU soft law? An argumentative perspective on the European Commission's soft law instruments in times of crisis. Theory Pract Legis 10(1):22–47

Daly P (2023) How binding are binding guidelines? An analytical framework. Can Public Adm 66:211–229

[40] Op.cit 31, para. 54.

Guillem Carrau J (2021) Denominaciones geográficas de calidad (Quality Geographical indications). Tirant lo Blanch, Valencia

Guillem Carrau J (2023) Towards the extension of geographical indication protection of the European Union beyond the agricultural products: an EU craft GI's system. EIPR 45(12):725–736

Kur A (2023) No strings attached to GIs? About a blind spot in the (academic) discourse on limitations and fundamental rights. IIC 54:87–94

Maroño Gargallo MM (2002) La protección jurídica de las denominaciones de origen en los derechos español y comunitario (The legal protection of the designations of origin under EU and Spanish law). Marcial Pons, Madrid

Mattera A (1991) El Mercado Único. Sus reglas. Su funcionamiento (The single market: its rules. Its functioning). Madrid

Megarry R (1944) Administrative quasi-legislation. Law Q Rev 60:125–129

O'Connor B (2004) The law of geographical indications. Cameron May, London

Rawlings R (2015) Soft law never dies. In: Elliott M, Feldman D (eds) The Cambridge companion to public law. Cambridge University Press, Cambridge

The Geographical Indications Coexistence with Trade Marks

7

Abstract

This chapter explores the coexistence of GIs and trade marks. It examines the decisions rendered by the EUIPO Board of Appeal on GIs, particularly in relation to the absolute ground of refusal under Article 7(1)(j) EUTMR. It also deals with the role of GIs in inter partes proceedings, analyzing a series of decisions, including *Campo Verde, Toro de Piedra*, and *Chenine Brut Premium Cava* Bottle, *Colombiano Coffee House*, *Romeral de Jabugo*, *Marqués de la Mancha* and *Padrón Auténtico*. Through these cases, it sheds light on the scope and extension of GI protection in trade mark disputes, providing valuable insights into the complex interactions between these two intellectual property rights. Some procedural situations related to GIs in EUIPO Appeal proceedings have been described as well in this chapter. In addition, the examination of the *Darjeeling tea* case will show how the interplay with collective trade marks can weak the GI. This part ends with an attempt to explain the differences with collective and certification trade marks.

The coexistence between trade marks and GIs, as a matter of fact, is admitted by the EU legislator and at multilateral level. The juxtaposition of the terms 'trade mark' and 'geographical indications' incites certain interest but somehow the understanding of an environment of coexistence more or less pacific.[1] Indeed, the frontiers between the two schemes have been opened and both type of signs are capable of dissemination of confidence, warrant and optimisation.[2]

GIs are, just like trade marks, signs used for the purpose of distinguishing goods. Unlike trade marks, GIs do not identify the commercial origin of a product from a particular undertaking, but its geographical origin from a specific place or region

[1] Pérez Ferreras (2003), p. 14.

[2] Audier (2003).

© The Author(s), under exclusive license to Springer Nature Switzerland AG 2026

J. Guillem Carrau, *Landmark EU Geographical Indications Case Law*, Law for Professionals, https://doi.org/10.1007/978-3-032-26101-4_7

and, connected therewith, in most cases a given quality or other characteristic of the product. In contrast to trade marks, GIs are not individual rights but indications that belong to all local producers of the relevant region. The right to use a GI is a right of all local producers which, in principle, cannot be withdrawn[3] when they produce according to the requirements, but the GI cannot be subject of appropriation, neither individually nor collectively.[4]

In fact, GIs have a difficult insertion within the classic intangible rights and, nevertheless, it aggregates elements from those intangible rights in order to create a singular concept within the orbit of the Commercial Law and, in particular, the Industrial Property. This singular situation highlights the need of improvement of conflict rules.[5]

In an international context, the WIPO documents that competing claims to the right in a given sign, either as a trademark, or as a GI, can be looked at from at least two different angles, namely from the point of view of trademark law, or the point of view of the law on GIs. The former approach may be best described by the question: "Can a particular sign constitute a valid trademark in a given territory, if, at the same time, it is claimed that this sign constitutes, in this territory, a GI?" The question that may characterize the latter approach would be: "Can a particular sign constitute a GI in a given territory if, at the same time, it is claimed that this sign constitutes, in that territory, a trademark?" Due to the variety of legal concepts existing in the field of GIs, it is more complex to illustrate the latter approach.[6]

As it is known, trade marks must be neither descriptive nor deceptive. Consequently, trade marks that consist of or contain a GI cannot be protected if use of such trade marks would be misleading as to the true origin of the products on which the trademark is used. Laws on trademarks specifically exclude from registration geographical terms that can be understood to constitute a reference to the origin of the relevant goods. This exclusion from registration usually depends on an assessment whether a geographical term used as a trade mark would be perceived by the public to indicate a connection between the origin of the goods and the trademark. However, it is possible to use geographical terms as trade marks if those terms are arbitrary such as, for example, "Antarctica" for bananas,[7] because that mark would not be understood to refer to the origin of the goods on which it is used. Moreover, it may be possible to use a geographical term as trade mark in cases where that trade mark, despite being originally descriptive, has acquired a distinctive character (or secondary meaning) through use.

On the top of that, in some territories, collective and/or certification trade marks have been a common resource to protect a GI term.[8]

---

[3] Knaak (2015), p. 843.

[4] Maroño Gargallo (2002), p. 165.

[5] Guillem Carrau (2021), p. 709.

[6] WIPO: Document SCT/6/3 Rev. on Geographical Indications: Historical Background, Nature of Rights, Existing Systems for Protection and Obtaining Protection in other Countries, page 26.

[7] Example taken from WIPO publication 760, page 166.

[8] Guillem Carrau (2021), p. 138.

In the field of the GIs, the conflicts are managed in a different way as, depending on the territories, the protection is deployed in several layers from unfair competition, passing off and protection as collective or certification trade marks to the protection as a *sui generis* right.[9]

7.1 A Balance Protection Between GIs Law and Trade Mark Law

When solving a conflict between trade marks, exclusivity principle applies and, therefore, the solution comes from the implementation of the principles of registry priority, speciality and territoriality. Then, in conflicts between GIs and trade marks, the mentioned conflict rules cannot be applied,[10] as such.

To provide an equal level of protection to GIs and trademarks, INTA and OIV are in favour of applying the priority principle to the conflicts between trade marks and GIs.[11]

Some scholars reflected on the limits of the priority principle in this type of conflicts, for instance, when the GI is internationally reputed[12] and, at the end, as the EU Court pointed out in *Gorgonzola* case law, the conflict must be solved by a national court.[13] In case of pacific coexistence, when the conflict deals with the respect of the exclusive right of use of a GI, the principles to be applied should be the same that are applied when, in trade mark law, a likelihood of confusion overall assessment is done; likewise it should be applied to GI protection the prohibition of taking unfair advantage of its reputation. Register is a key point at this stage to implement the veracity principle on the GI's territorial extension and characteristics.[14]

Therefore, the exclusivity principle plays a crucial role, and the legislator in some jurisdictions has already taken a position in this context. In this respect, trademark legislation tends to specifically exclude the registration of geographical terms that could be considered a reference to the origin of the products in question, without prejudice to what has been stated regarding geographical collective marks and geographical guarantee marks, where registration is permitted, but with specific limitations regarding the right to prohibit and the freedom to establish the rules governing the use of the trademark.

Coexistence is typically declared, as EUTMs and GIs are often viewed as complementary rights. This complementary relationship is rooted in the fact that they

[9] WIPO: Document SCT/6/3 Rev. on Geographical Indications: Historical Background, Nature of Rights, Existing Systems for Protection and Obtaining Protection in other Countries, page 27.

[10] Reichert (2003), p. 20.

[11] INTA (1997) Protection of Geographical Indications, available at https://www.inta.org/wp-content/uploads/public-files/advocacy/board-resolutions/Protection-of-Geographical-Indications-11.2019.pdf OIV (1994) RESOLUTION ECO 3/94 www.oiv.org.

[12] Bercovitz Rodriguez-Cano (2000), pp. 94 y 95; Reichert (2003), p. 20; Grazioli (2003), p. 37 and Stern (2003), p. 3.

[13] C-87/97 - Consorzio per la tutela del formaggio Gorgonzola, ECLI:EU:C:1999:115.

[14] Stern (2003), p. 44.

protect different aspects of intellectual property. GIs have a unique, cross-cutting position within EU intellectual property law.

The various existing legal traditions on the protection of geographical indications of provenance were surpassed by European law through the creation, via Regulation, of two clearly defined legal categories, PDO and PGI, with identical scope of protection; the establishment of a European Register for these categories and regulated procedures for access to it; as well as the development of management bodies and control structures that bring homogeneity and security to the system.

By Regulation (EEC) 2081/92, the European legislator connected the existence of PDOs and PGIs with trade marks, and incorporated rules to facilitate coexistence and resolve potential conflicts that may arise between these two types of distinctive signs.[15]

In the EU, the GIs legal framework on the one hand and EUTMR[16] on the other, contain provisions to define the rules of coexistence of GIs and EUTM. On one side, according to Article 30 Regulation (EU) 2024/1143, a name must not be registered as a GI where, in the light of a trade mark's reputation and renown and the length of time that it has been used, registration of the name proposed as a GI would be liable to mislead the consumer as to the true identity of the product. Likewise, according to Article 31 Regulation (EU) 2024/1143, an application for the registration of a trade mark the use of which would contravene Article 26 Regulation (EU) 2024/1143 must be rejected if the application for registration of the trade mark is submitted after the date of submission to the Commission of the application for the registration of the GI and must be declared invalid if registered in breach of this rule. However, in some cases, the use of the GI, once registered, and that of the relevant trade mark shall be permitted when it has been applied for, registered, or established through use in good faith in the EU, where that possibility is provided for in the relevant legislation, before the date of submission of the application for registration of the GI to the Commission. On the other side, Article 7(1)(j) EUTMR provides that '*1. The following shall not be registered:....(j) trade marks which are excluded from registration, pursuant to Union legislation or national law or to international agreements to which the Union or the Member State concerned is party, providing for protection of PDO and GIs shall not be registered'.* This absolute ground of refusal can also be invoked in the context of cancellation proceedings, pursuant to Article 59(1)(a) EUTMR and it can work only when the goods and services applied for are comparable to the goods under GI protection. However, this objection can be overturned when the goods and services applied for are described as following the GI specification (or goods in which the GI is a relevant ingredient). Following the new version EUIPO Guidelines that entried in force in July 2026, the restriction is not possible when the trade mark 'evokes' the GI for any good or service and

[15] Reichert (2003), p. 20.

[16] Regulation (EU) 2017/1001 of the European Parliament and of the Council of 14 June 2017 on the European Union trade mark (OJ L 154, 16.6.2017, p. 1).

when the trade mark 'uses or evokes' the GI for processed or manufactured goods where GI product would be an ingredient, part or component; so the objection based on Article 7(1)(j) EUTMR cannot be waived. On the contrary, the restriction is possible when the trade mark uses the GI for identical good and related services if it is restricted to goods complying with the GI requirements.

The relationship between trade marks and GIs should be clarified with regard to criteria for the rejection of trade mark applications, the invalidation of trade marks and the coexistence between trade marks and GIs.[17]

Likewise, the guarantee or certification marks referred to in Article 28(4) of Directive (EU) 2015/2436 and collective marks referred to in Article 29(3) of that Directive, as well as collective marks as referred to in EUTMR can be used on labels together with the GI. In the EU, there were cases where GIs coexist with organic production certification within the same product.[18]

An example of the coexistence of trade mark and GIs is the term Bavaria, which was subject matter of a well-known litigation identified with the same name. The referring Court wanted to know whether the GI 'Bayerisches Bier' ought not to have been refused in the light of 'the reputation and renown' of the Bavaria trade marks and the length of time for which they have been used, and if the indication 'Bayerisches Bier' was 'liable to mislead the consumer as to the true identity of the product'. In that sense, the finding to the effect that the GI 'Bayerisches Bier' and the trade mark 'Bavaria' were not in the situation which the applicable at that time Article 14(3) of Regulation No 2081/92 was designed to cover appeared not to be incorrect or to go beyond the discretion of the Commission and the Council in determining that issue, given that the higher the reputation or renown of a mark, and thus the stronger its distinctiveness, the lower would be the likelihood that consumers might be misled to the extent of ascribing a product with a given GI to that trade mark.

Therefore, the Court concluded that the situation was to be interpreted as not affecting the validity and the use of pre-existing trade marks including the term Bavaria provided, however, that those marks were registered in good faith and are not liable '*to be invalid or revoked on the basis of First Council Directive 89/104/EEC of 21 December 1988 to approximate the laws of the Member States relating to trade marks as provided for under Article 14(2) of Regulation No 2081/92, which it is for the national court to determine*'.[19]

---

[17] See Recital 39 Regulation (EU) 2024/1143.

[18] Commission Implementing Regulation (EU) No 268/2013 of 18 March 2013 approving non-minor amendments to the specification for a name entered in the register of protected designations of origin and protected geographical indications (Oberpfälzer Karpfen (PGI) OJ L 82, 22/03/2013, pp. 43–44.

[19] C-343/07, Bayerisches Bier ECLI:EU:C:2009:415, para 133.

Thus, the protection of GIs and DOs is politically desired, and the EU Courts show a clear tendency to decide in favor of the protection of GIs and DOs when applying the requirements for the registration of GIs and DOPs.[20]

7.2 EUIPO Guidelines: An EU *Soft Law* in the Context of the Coexistence of Trade Mark and GIs

EUIPO current practice is reflected in a series of Guidelines that are intended to be of practical use both to Office staff in charge of the various procedures and to users of the Office's services. These guidelines are revised and are the main point of reference for users of the European Union trade mark system, and for professional advisers who want to ensure they have the latest information on EUIPO examination practices.

As regards GIs, the practice of the Office in Part B, Examination in relation to Article 7(1)(j) EUTMR, which is the absolute ground of refusal that contemplates the GIs is condensed in Chapter 10; and, in the part C, Opposition, where section 6 contains the practice of the office as regards Article 8(6) EUTMR. In July 2026, a revised version of the EUIPO Guidelines entered into force. Both relevant chapters were redrafted to clarify the expanded scope of ex officio examination under Article 7(1)(j) EUTMR. This change reflects Article 31 of Regulation (EU) 2024/1143, which removes the previous limitation of ex officio examinations to identical or comparable goods. Furthermore, the Guidelines now incorporate clarifications from the *Nero Champagne* judgment regarding the Office's duty to act. The EUIPO will now refuse an application if a trademark evokes a GI for any goods or services. This also applies where a trade mark uses or evokes a GI for processed products containing the GI product as an ingredient or component. Regarding the distinction between "use" and "evocation", EUIPO explains that 'use' generally applies when the GI name is included in the sign but is perceived as a separate and independent component; and 'evocation' now encompasses several specific scenarios: first, the use of the GI in an adjectival form (previously categorized as "use"); second, the inclusion of non-generic parts of compound GIs within the trade mark; and a third one, where trade marks where the registered GI name interacts with other elements, unless the name constitutes a separate, logical conceptual unit that diverges from the perception of the GI. In this final instance, no conflict is deemed to exist. Another change is the introduction of the concept of 'notiority': '*a GI may, however, because of its longstanding presence on the market, marketing efforts and commercial success, acquire notoriety beyond the intrinsic baseline reputation accorded to all GIs which will increase the potential for exploitation, weakening, dilution or detriment to its reputation*' .[21]

[20] Günzel (2012), p. 293.

[21] See EUIPO Trade mark guidelines, Chapter 10 Trade marks in conflict with geographical indications (Article 7(1)(j) EUTMR).

Guidelines are shaped as an instrument of *soft law*.[22] In the case of EUIPO Guidelines, they have become potentially a high persuasive tool because of their arguments, which is an instrument of rational persuasion as, in the case at hand, they legitimise the EUIPO point of view and actions. They cover completely the practice of the Office and their 'explain and convince' approach looks works adequately in the context of the intellectual property rights. Its legal effects are defined by the approval by the EUIPO Executive Director after consultation with the Management Board in accordance with Article 153(1) (l) EUTMR.[23]

7.3 The Relevance and Persuasive Value of Decisions Rendered by the Boards of Appeal on GIs

An insight into the Decisions of the EUIPO Boards of Appeal on GIs helps to understand how the protection of the GIs as a specific intellectual property right weight gainer in many litigations because of the singular position is, is assigned to GIs by the European legislator in the IP field.[24]

There is an unlimited number of Decisions of the Boards of Appeal where the subject matter is related to GIs in a way or another and, as it is known, the Boards of Appeal are not bound for precedent Decisions, nor in the field of GIs.[25]

Specifically, the activity of the EUIPO Board of Appeal has contributed along the years with Decisions on matters that contain some quite fluid concepts, like evocation, where there is a large amount of case law and doctrine.[26]

First and foremost, a review of the recent Boards of Appeal's decisions on GI's allows to muse the concept of comparable goods, the restrictions of goods and services and when not to look for a EUTM with a GIs reference.

It is also part and parcel of our effort to pay attention to GIs protection behind the *inter partes* proceedings' scenes by scrutinizing decisions on the GIs scope of protection against direct and indirect use and reputation, misuse, evocation, and misleading indications. On top of that, there are some Board of Appeal Decisions on procedural aspects that will allow to identify some particularities in appeal proceedings related to GIs like a request for GIs external expertise, the substantiation of the GIs' right or the right to be heard.

[22] See Chapter 5.

[23] The Guidelines on EU trade marks and the Guidelines on registered Community designs that are currently in force were adopted by the Executive Director on 30/04/2025 (Decision No EX-25-3), and entered into force on 01/05/2025.

[24] Guillem Carrau (2024), p. 749.

[25] Marchisio (2020), p. 893.

[26] Zafrilla Diaz-Marta and Kyrylenko (2021), p. 442.

7.4 Exclusion of Registration as Trade Marks: The Absolute Ground of Refusal (Article 7(1)(j) EUTMR)

Article 7(1)(j) EUTMR provides that "*1. The following shall not be registered:.... (j) trade marks which are excluded from registration, pursuant to Union legislation or national law or to international agreements to which the Union or the Member State concerned is party, providing for protection of PDO and GIs shall not be registered*".

This absolute ground of refusal can also be invoked in the context of cancellation proceedings, pursuant to Article 59(1)(a) EUTMR and it can work only when the goods and services applied for are comparable to the goods under GI protection. However, this objection can be overturned when the goods are identical and services related applied for are restricted as described below, as following the GI specification (or goods in which the GI is a relevant ingredient)

The EUIPO Board of Appeal Consistency report on 'Evocation of geographical indications in absolute grounds' outlines, the uniformity and consistency of their Decisions on appeals with any GI element is quite remarkable, as also shown by the fact that appeal decisions on absolute grounds for refusal are not often challenged before the Court and that, if challenged, they are usually confirmed.[27]

7.4.1 Comparable Goods: *BNIC Cognac Case* Criteria

A EUTM application could not be refused under Article 7(1)(j) EUTMR if the goods and services applied for were not comparable to those protected by the GI. The Office did not raise objections ex officio against different goods. As said in points 7.1 and 7.2 of this chapter, the 2026 version of the EUIPO Guidelines would imply that the restriction is not possible when the trade mark 'evokes' the GI for any good or service and when the trade mark 'uses or evokes' the GI for processed or manufactured goods where GI product would be an ingredient, part or component; so, as said before, the objection based on Article 7(1)(j) EUTMR cannot be waived.

The notion of comparability of products is relevant for both absolute grounds, under Articles 7(1)(j) and (k) EUTMR, and relative grounds under Article 8(6) EUTMR. The notion of 'comparable products', as referred in chapter 3, was developed in *BNIC*[28] (also known as *Cognac case*) and it is essential to understand the protection conferred to GIs because the Court identified the following non-exclusive list of factors:

a) Whether the products have common objective characteristics, such as method of elaboration, physical appearance, or use of the same raw material,

[27] Board of Appeal (2023): Consistency report on 'Evocation of geographical indications in absolute grounds' Consistency Circle Absolute Grounds June 2023 available at https://euipo.europa.eu/tunnel-web/secure/webdav/guest/document_library/contentPdfs/about_euipo/Board_of_appeal/research_reports/Evocation-of-geographical-indications-in-absolute-grounds_en.pdf, p. 15.

[28] C-4/10 & C-27/10, *BNIC Cognac*, EU:C:2011:484, para. 54.

b) Which are the circumstances under which the products are consumed, from the point of view of the relevant public,
c) Whether they are frequently distributed through the same channels, and
d) Whether they subject to similar marketing rules.[29]

In addition, the comparability of products is assessed from the perspective of the average consumer. This is an average European consumer who is reasonably well informed and reasonably observant and circumspect.[30]

In *Manchego*,[31] the Court said that it must be understood as covering European consumers, including consumers of the Member State in which the product giving rise to evocation of the protected name is made or with which that name is geographically associated and in which the product is mainly consumed.

Following the BNIC criteria, the EUIPO Boards of Appeal have considered some straightforward cases of comparable products, where the contested sign was registered for goods of the same genus as those protected by the PDO or PGI; for instance, like alcoholic beverages (except beer) and wine; meat or meat-based products containing ham and ham; or liqueur products and vodka.

An EUIPO Boards of Appeal Consistency report[32] outlines that the decision-making practice in comparable goods has been built up in consistency with the case-law of the EU courts and within the Boards. Very few cases are appealed before the General Court, and they are usually endorsing the Board of Appeal's Decisions. The same may be said in relation to the guidelines; the decisions of the Board do not diverge from the practice laid down in the EUIPO guidelines. However, the Guidelines say that comparability should be interpreted restrictively, whereas the case law of the Board may be considered to take a slightly broader approach.

For instance, in *Fratelli Parmigiani*,[33] the EUTM had been applied for meat-based products which could potentially be comprised of or contain ham or some other meat that is comparable to the protected Parma ham. Although *Prosciutto di Parma* was specifically excluded from the amended specification, it was enough that goods applied for could be made from a product comparable to the protected PGI to justify the ground for refusal. The Board of Appeal decided that those meat based products ('meat, preparations for stuffing pasta made from meat, meat extracts; cooked, seasoned, smoked charcuterie of all kind of meat; sausages, wurstel, spread-

[29] Recital 34 Regulation (EU) 2024/1143 declares that to establish whether products are comparable to the products designated by a GI, account should be taken of all relevant factors. Those factors should include whether the products have common objective characteristics, such as method of production, physical appearance or use of the same raw material; under which circumstances the products are utilised from the point of view of the relevant public; whether they are frequently distributed through the same channels; and whether they are subject to similar marketing rules.

[30] From Gorgonzola (04/03/1999, C-87/97, *Cambozola*, EU:C:1999:115, para. 25) to the Scotch Whisky Association (22/02/2018, C-44/17, *The Scotch Whisky Association*, EU:C:2018:111, para. 56).

[31] C-614/17, ECLI:EU:C:2019:344, para. 50.

[32] Op.cit. 29.

[33] 16/01/2013, R 659/2012-5, *Fratelli Parmigiani Alta Salumeria Salame di Parma*, para. 11 and 14.

able pâté of any kind and of any flavour, pâté made from meat, meat products) should not have been allowed registration despite the fact that the applicant had attempted to limit its list of goods to those complying with the specification of the PGI by adding "but not including meat-based products entitled to the geographic indication *Prosciutto di Parma*' to the end of its list.

In *Parmatutto*,[34] the Board found that foodstuffs based on meat in Class 29 were all likely to refer to or include as their main ingredient, ham and shoulder of pork, both of which are both products falling under the specification protected by the PDO *Prosciutto di Parma* and by the PGI *Coppa di Parma.*

As regards *Vana Tallinn Authentic Estonian Liqueur,*[35] the Board of Appeal rejected the registration of a figurative sign featuring the designation Estonian Liquer for 'Liqueurs; liqueur-based cocktails and mixed drinks' because of the protection afforded to the PGI *Estonian Vodka* for vodka. In the Board's opinion it was reasonable to hold that liqueur products were comparable to the vodka. The goods applied for covered drinks which had common objective characteristics, such as their type (spirit drinks), method of elaboration (distillation), physical appearance and use of similar raw (agricultural) materials. Moreover, they were consumed on similar occasions (namely on special occasions, for enjoyment and relaxation), and they were also frequently distributed through the same channels (such as alcohol shops or alcohol sections in supermarkets) and subject to similar marketing rules.

When assessing comparability in *Colombiano Coffee House*,[36] the fact was that the contested sign was applied for goods in Class 30 'Tea, cocoa, sugar, rice, tapioca, sago; flour and preparations made from cereals, bread, pastry awnd confectionery, ices; honey, treacle; yeast, baking-powder; salt, mustard; vinegar, sauces; spices; ice' and services in Class 43 'Services for providing food and drink; temporary accommodation'. The request was based on the PGI *Café de Colombia*, registered for coffee. On the contrary, the goods in Class 30, apart from tea and cocoa, could not be comparable to coffee and the contested sign's services in Class 43 could not be comparable products, since their link with coffee was vague and imprecise. As for tea and cocoa, which the Board deemed not to be comparable to coffee, the Board recognized that they had certain elements in common with coffee, such as their physical appearance as well as their method of use (all can be in soluble powder form to be mixed with boiling water in order to make a hot drink). Nevertheless, the Board concluded that these common features would not justify the inclusion of these products in the same class for the purposes of protection of a PGI as tea and cocoa are products having their own methods and tradition of cultivation, harvesting and production, and the peculiar characteristics and methods of the PGI '*Café de Colombia*' could not be deemed to be applicable thereto. This reference to "the same class for the purposes of protection of a PGI" is problematic, as the term class is not defined under GI law, is a term of art in trade mark law which clearly does not apply in the assessment of comparability, and seems to indirectly introduce a new,

[34] 20/01/2014, R 1900/2013-5, *Parmatutto*, para. 21, 22.

[35] 02/12/2013, R 1299/2013-1, *Vana Tallinn Authentic Estonian Liqueur*. para. 26.

[36] 27/03/2014, R 1200/2013-5, *Colombiano Coffee House*, para. 31.

potentially relevant criteria for the assessment of comparability. In the absence of further case law relying on this criterion, this statement should be taken as obiter, in particular in the light of the Board's subsequent reasoning based on the BNIC criteria: whether they are frequently distributed through the same channels and whether they subject to similar marketing rules.

In *Toro,*[37] the Board of Appeal found fruit juices and beverages to not be comparable with wine. They are not comparable under the BNIC criteria as they undergo clearly different production processes, are on sale in different sections of a store and are not subject to the same or to similar rules covering preparation or marketing. In *Toroloco (fig)*[38] the Board considered that it would be used for goods which were comparable to the products protected under the GI but which were not covered by the registration.

In *Yorkshire Provender,*[39] the Board reasoned that cheese is a dairy food made from pressed milk and that the applied for milk and dairy products and dairy desserts clearly include cheeses or can have cheese as a constituent product.

The Board of Appal in its *Ibiza Flirt*[40] decision, considered that the terms "alcoholic beverages" and "spirits" in a specification of goods would encompass any kind of spirits and thus also the relevant aniseed-flavoured spirit drinks protected by the PGI *Hierbas Ibicencas*.

In the case of *Portobello Road No 171*,[41] the Board held that Gin is not a comparable product to wine.

In *Quevedo Port*,[42] the contested goods 'olive oils' were considered not comparable to wine, and in *PortSoy*,[43] the contested 'Scotch Whisky complying with the specifications of the PGI *Scotch Whisky*' was considered not comparable to wine and noted that whisky was not a grape wine product under EU Law. Both decisions have been endorsed by the General Court.[44]

[37] 23/02/2016, R 612/2015-4, *Toro*, para. 14, 15. To the contrary for Toro, see 03/09/2020, R 2751/2019-4, TORO DE PIEDRA (fig.) / Toro denominación de origen et. Al.

[38] 18/11/2014, R 2462/2013-2 – TOROLOCO (FIG. MARK), para. 20.

[39] 25/01/2018, R 674/2017-1, *Yorkshire Provender*, para. 39–42.

[40] 01/06/2016, R 2531/2015-2, *Ibiza Flirt*, para. 64.

[41] 15/06/2016, R-1105/2015-4, *Portobello Road No 171*, para. 20.

[42] 14/11/2023, R 2471/2022-4, Quevedo Port / Port, para. 31.

[43] 14/11/2023, R 1885/2022-4, PORTSOY / Port, para. 35. In this case, the Board followed the interpretation done by the Opposition Division. See para. 33 '*The Opposition Division correctly reasoned that there are significant differences between the features of the goods under comparison, in terms of, inter alia, ingredients, alcohol content and taste, of which the average consumer is well aware. Whisky usually has more than 40% alcohol content, while wine protected by the PDO 'Port' has between 17% and 21%. Whisky uses malted barley as its main ingredient, while wine is produced from grapes. Furthermore, the methods of production differ since wine is produced by way of fermentation, while whisky is produced by distillation. Whisky is generally not served in the course of a lunch or a dinner, whereas wine can be paired with many different kinds of food*'.

[44] T-23/24 - IVDP v EUIPO - Vinoquel-Vinhos Oscar Quevedo (Quevedo Port) ECLI:EU:T:2025:182, para.; and T-40/24 - IVDP v EUIPO - The Benriach Distillery Company (PORTSOY) ECLI:EU:T:2025:183.

In both cases, the Board of Appeal considered that the contested goods were not comparable to ‘wine’ protected by the PDO *Porto*. In addition, the EUIPO Board of Appeal discarded the existence of any evocation and exploitation of the reputation because the PDO *Porto* failed to adduce any evidence to demonstrate it. Both cases were appealed and the General Court had another opportunity to entail the analysis of comparability criteria but this has not happened as the judgment was focus in the exam of the existence of a commercial use for non-comparable products and it followed the conclusion of the Board on the average degree of similarity.

In *Portsoy*, contrary to what the applicant claimed, the Court considered that it was not necessary to find that the signs at issue were visually and phonetically similar to a particularly high degree. In the light of the differences and similarities between the signs noted in the contested decision, it was appropriate to find that the Board of Appeal was entitled to conclude that the signs were not visually or phonetically similar to a particularly high degree, let alone close to identical. Even though the beginning of the contested sign, namely the element ‘port’, fully reproduced the earlier PDO, that initial part was coupled with another part, the element ‘soy’, specific to the contested sign, which had to be taken in account both visually and phonetically. It has been criticised by the PDO *Porto* that the Board of Appeal ignored certain factors referred in BNIC case law, such as the methods of production; ignored other factors put forward by the applicant, such as the nature of the products (alcoholic beverages), the similar prices, and the existence of competition between whisky and port wine as regards the consumers targeted; and misunderstood the occasions on which whisky and port wine are consumed. In that context, it was argued that the Board of Appeal might not limit itself to comparing the products in the light of their raw material, grapes or grain, or the provisions which protect their designations of origin, relating to grapevine products or spirit drinks. In their view, nothing in Article 103(2)(a)(i) of Regulation No 1308/2013 supports the conclusion that ‘comparable products’ is a category restricted to wines.[45]

In *Quevedo Port*, in the context of examining the existence of a commercial use of the name of the protected by the earlier PDO, the Court, by quoting *Champanillo* case[46] and discarding the example of the previous decision on *Nero Champagne*,[47] affirms that, when the products are not comparable, there is ‘use’ when the degree of similarity between the signs is particularly high and close to identicality in such a way that the protected name is used in a form that cannot be dissociated from it. The General Court also recalled *Champanillo* case to refer that the concept of ‘use’ must be interpreted strictly to ease the differentiation from the concept of ‘evoca-

[45] See arguments of the appellant as reproduced by the Court in T-40/24 - IVDP v EUIPO - The Benriach Distillery Company (PORTSOY) ECLI:EU:T:2025:183, para. 53.

[46] C-783/19, Comité Interprofessionnel du Vin de Champagne, EU:C:2021:713, para. 38 and 40.

[47] See para. 46 ‘*Accordingly, the legality of the decisions of the Boards of Appeal must be assessed solely on the basis of that regulation as interpreted by the EU judicature and not on the basis of a previous decision-making practice (see, to that effect, judgment of 26 April 2007, Alcon v OHIM, C-412/05 P, EU:C:2007:252, paragraph 65 and the case-law cited)*’; and as regards the Champagne Nero Decision: 17/02/2023, R 531/2022-2, NERO CHAMPAGNE, para. 21.

tion'. In those circumstances, the signs at issue were visually and phonetically similar to an average degree,[48] for the Court it was not necessary to find out whether the signs at issue were visually and phonetically similar to a particular high degree, as the Porto PDO representative argued.

In *Port Charlotte*, the GC rejected a similar complain because there was no need, at that stage, to decide whether whisky and liqueur wines, such as port, were comparable.[49]

> 64 In this connection, it should be noted that the Board of Appeal stated, in essence, that the geographical indications within the meaning of that provision were protected only for wines and, therefore, for goods that were neither identical nor comparable to a product denominated 'whisky', namely a spirit drink with a different appearance and degree of alcohol that cannot comply with the product specification for a wine (paragraph 20 of the contested decision).
>
> 65 That assessment is not vitiated by error, given that a spirit, the product of the alcoholic fermentation of grains, such as whisky, is, by definition, incapable of satisfying the conditions of a product specification for a wine, the product of the alcoholic fermentation of grapes, within the meaning of Article 118m(2)(a)(i) of Regulation No 491/2009, read in conjunction with Article 118c(2) thereof.

On the contrary in *Portwo*,[50] the Board of Appeal indicated that the mark applied for and the PDO '*Porto*' were, visually and aurally, very similar and that that use of the mark applied for in relation to the spirits in Class 33 was likely to affect consumers' preferences, by inducing them to believe that those spirits complied with the strict requirements and quality standards of the PDO 'Porto' enjoyed an exceptional reputation and appreciation among the relevant public, for its qualities and its characteristics. The General Court endorsed the Board of Appeal view on the comparability of wine and spirits.[51]

In *Porte Noire*,[52] the contested goods 'brandy and distilled spirits' were considered similar to wine while whisky and wine were not similar, even though whisky is also a distilled spirit. The Board of Appeal concluded evocation have been fulfilled only in relation to the contested *alcoholic beverages, except beers; sparkling wines; wine; alcoholic wines; spirits and liqueurs; alcopops; prepared alcoholic cocktails; alcoholic beverages containing fruit; alcoholic bitters; alcoholic cocktail mixes; alcoholic cocktails in the form of chilled gelatins; alcoholic coffee-based beverage; alcoholic energy drinks; alcoholic punches; alcoholic tea-based beverage; aperitifs; brandy; coffee-based liqueurs; cream liqueurs; distilled spirits; fortified wines; grappa; pre-mixed alcoholic beverages, other than beer-based; prepared wine cocktails; schnapps; sherry; vermouth* in Class 33. The opposition failed, and the

---

[48] See para. 32 to 35 of the contested decision.

[49] T-659/14, Instituto dos Vinhos do Douro e do Porto v OHMI - Bruichladdich Distillery (PORT CHARLOTTE), ECLI:EU:T:2015:863, para. 67.

[50] 21/04/2020, R 993/2019-2, Portwo gin / Porto, para 53.

[51] T-417/20 - Esteves Lopes Granja v EUIPO - IVDP (PORTWO GIN), ECLI:EU:T:2021:663, para. 51.

[52] 14/11/2023, R 1019/2023-4, PORTE NOIRE / Port, para. 70 to 73.

appeal was dismissed, for the remaining contested goods, namely *herb liqueurs; absinthe; whisky; blended whisky; bourbon whiskey; malt whisky; calvados; dry cider; gin; kirsch; rum; vodka*; the relevant consumers, English- and French-speaking or not, will not establish a link between the contested sign, 'PORTE NOIRE' and the PDO *Porto* due to the significant differences in their features. As regards this last list of goods, the Board stated that the opponent failed to provide evidence and/or at least develop a coherent line of argument to demonstrate the link between the contested sign used to designate these non-comparable and rather distinct goods and the earlier PDO *'Porto'/ 'Port'*.

Pursuant to a series of decisions rendered by the Board of Appeal, a *BNIC* approach has been established, which sets forth a framework for the comparability of the goods. Some decisions show a conception of 'comparable goods' which has been considered extremely narrow within the aims and rules protecting GIs in the EU.[53]

7.4.2 Evocation of a GI as an Absolute Ground Refusal

Evocation as an absolute ground of refusal under Article 7(1)(j) EUTMR means that the mark applied for might be led to make a mental connection between the GI product and origin of the one marketed under the applicant's trade mark.

In other words, it must be established, based on aural and visual similarities, whether these are sufficient to evoke the product protected by the GI. These factors have been systematically identified by the Board of Appeal in cases like *Yorkshire Forced Rhubarb / Yorkshire Provender*[54] (same initial word element), *Val di Magra / Ca' del Magro*[55] (very similar final word element), *Colli di Imola / Mappa di Imola*[56] (same final word element), and *Fara / La Farra*[57] (very similar word element), where the visual and phonetic similarities leading to evocation typically derive from the partial incorporation of one or more elements of the protected GI in one of the verbal elements of the contested mark. On top of that, the fact that the applicant had no intention to evocate the GI and that there is no risk of confusion is irrelevant. In any event, the alleged peaceful coexistence in the EU has not been proven.

Similar decisions were taken in *Casa Timis*,[58] where the rejection was based on the fact that the application was evoking the PGI 'Viile Timişului', and *in Mehr*

[53] See the arguments of the appellants as reproduced by the Court in T-40/24 - IVDP v EUIPO - The Benriach Distillery Company (PORTSOY) ECLI:EU:T:2025:183, para. 51.

[54] 25/01/2018, R 674/2017-1, YORKSHIRE PROVENDER, para. 43 to 50.

[55] 29/09/2017, R 1006/2017-5, CA' DEL MAGRO, para. 33 (CA' DEL MAGRO / VAL DI MAGRA).

[56] 19/09/2019, R 1223/2019-5, Mappa di Imola, para. 17.

[57] 13/05/2020, R 2230/2017-2, La farra, para. 20.

[58] 20/01/2020, R 1315/2019-5, CASA TIMIŞ (fig.), para 31.

allgaü past in keinen kässe[59] in which the refusal happened as the claimed goods were identical to the products covered by four PDOs registered the PDOs ('Allgäuer Sennalpkäse', 'Allgäuer Weisslacker', 'Allgäuer Emmentaler', 'Allgäuer Bergkäse') prior to the filing date of the EUTM application.

All in all, it is sufficient, for Article 7(1)(j) EUTMR to apply, to find evocation by reference to consumers in part of the Union, even in the territory of a single Member State.[60]

To sum up on this point, the Board of Appeal produced a research report on the matter,[61] which listed a set of irrelevant factors for evocation. Among them, we can refer, for instance, whether the true origin of the product is indicated in the contested sign or whether it contains the geographical indication accompanied by 'delocalising' expressions, such as 'style', 'type', 'method', 'as produced in, 'imitation', 'flavour', 'like' or any other similar term; whether the name that benefits from a registered GI protection is unknown to the general public or relevant public or has several meanings; whether the mark for which registration is sought may mislead the public or whether it entails a likelihood of confusion for that public concerning the origin of the goods; whether the bodies responsible for protecting the relevant designation of origin did not oppose the registration of the mark; the possible existence of a family of marks, the contested mark being one of that family; the reputation of the contested trade mark and the market practice concerning the contested goods.

The Board of Appeal consistency report on 'Evocation of geographical indications in absolute grounds' concludes that '*The case-law analysis has shown that the Board' decision-making practice is consistent with the case-law of the EU courts and within the Board. The uniformity and consistency of the case-law is quite remarkable, as also shown by the fact that appeal decisions on absolute grounds for refusal are not often challenged before the Court and that, if challenged, they are usually confirmed'*.

7.4.3 Looking for a Safe Position: A Limitation Regarding GI's Goods and Services? A Lesson Learnt from *Nero Champagne*

According to EUIPO Guidelines, an EUTM application cannot be refused under Article 7(1)(j) EUTMR if the goods and services applied for are described as following the GI specification. Therefore, objections raised under Article 7(1)(j) EUTMR may be waived if the relevant goods are restricted to comply with the specifications of the GI in question. Since 2023. in the EUIPO Guidelines, it

[59] 08/04/2020, R 1920/2019-1, Mehr allgäu passt in keinen Käse, para. 40.

[60] 02/05/2019, C-614/17, Queso Manchego, EU:C:2019:344, para. 50 (ROCINANTE / QUESO MANCHEGO); 29/09/2017, R 1006/2017-5, CA' DEL MAGRO, para. 30 (CA' DEL MAGRO / VAL DI MAGRA.

[61] Op cit. 29.

was said that the Office considers that, in the context of examining absolute grounds for refusal, the protection of GIs under all EU regulations cannot extend to all possible products. The protection is limited to: goods identical to the product covered by the GI, including when such goods constitute the specific object of services such as retail, wholesale, import/export, provision of drink and food, production of [the product covered by the GI] for others; goods comparable to the product covered by the GI; and goods in which the GI is a relevant ingredient. As described before, the new EUIPO pratice implies that the limiation is not longer possible when the trade mark 'evokes' the GI for any good or service or when the trade mark 'uses or evokes' the GI for processed or manufactured goods where GI product would be an ingredient, part or component.[62]

In the EUIPO Board of Appeal, there is a consolidate approach in favour of allowing trade mark applications which contain a limitation for GIs goods and services. For instance, this happens when the EUTM applicant proposes a limitation as suggested by the examiner. In this sense, the Board of Appeal confirm EUTM applications with a limitation in the form of a clear, precise and unconditional limitation for GIs goods (*Olympia,*[63] *Fendente,*[64] *Allevatori bufalini campani abc caseificio del' 78 (fig.)*[65]) either for services (*Félix de murtiga Jabugo*[66]) or even when the GI term is included in a name to be registered as a colective EUTM (*cava de paraje calificado*[67]). In some cases, the appeal is closed following a restriction which overcomes the initial objection of the examiner (*The Great Earl Irish Whiskey*[68] *and Maison Sicile (fig.)*,[69] for instance).

Moreover, the limitation is not accepted by the Board of Appeal when it is not fully aligned to the one suggested by the examiner. This was what happened with

---

[62] See https://guidelines.euipo.europa.eu/2214311/2226841/trade-mark-guidelines/4-situations-covered-by-the-eu-regulations-and-absolute-grounds-examination.

[63] 18/01/2018, R 1229/2017-5, OPTIMA feinkost Olympia (fig.) para. 15.

[64] 20/11/2020, R 1450/2020-5, FENDENTE MAREMMA TOSCANA Le uve vengono raccolte a mano BURATTINI (fig.), para. 28.

[65] 04/03/2021, R 148/2021-5, allevatori bufalini campani abc caseificio dal' 78 (fig.), para. 15–17.

[66] 05/12/2023, R 1120/2023-2, FÉLIX de Múrtiga JABUGO (fig.) / FELIX et al., para. 21.

[67] 15/09/2017, R 500/2017-2, CAVA DE PARAJE CALIFICADO, para. 16. For the Board of Appeal the collective mark applied for, 'CAVA DE PARAJE CALIFICADO', was descriptive according to art. 7(1)(c) EUTMR. However, since the Board considers that it was a collective EUTM which described the geographic origin of the products covered by the mark in Classes 33, 35 and 39, the exception contained in art. 66(2) EUTMR for collective TM applied. 'CAVA' is a Denomination of origin indicative of the product's origin related to several areas within the Spanish territory, where several separate areas can be regarded as the same geographic origin. The other words 'DE', 'PARAJE' and 'CALIFICADO' are generic words which contribute to reinforce the mark's descriptiveness of the geographic origin. The Boards concluded that Art. 7(1)(b) did not apply either, given that geographical indications (including D.O.) have a fundamentally distinctive character. Furthermore, if the exception to art. 7(1)(c) EUTMR pursuant to art. 66(2) EUTMR applied, not applying it as well to art. 7(1)(b) EUTMR would have deprived art. 66(2) EUTMR of its function/object.

[68] 12/08/2020, R 468/2020-5, The Great Earl Irish Whiskey, para. 11–12.

[69] 29/03/2023, R 1667/2021-4, MAISON SICILE (fig.). para. 17.

'Mappa di Imola'. The sign 'Mappa di Imola' evoked the protected GI 'Colli d'Imola' for wines. The examiner proposed that the ground for refusal might be overcome by limiting the goods affected by the designation of origin in Class 33 as follows: wines complying with the specifications of the protected designation of origin 'Colli of Imola'. The applicant did a limitation of the list of the goods applied for to 'wines from Romagna except those belonging to the GI 'Colli d'Imola', which was exactly the opposite to the proposal of the examiner.[70]

In *Nero Champagne*,[71] the Board of Appeal found that use of the contested mark for some services would exploit the reputation of the PDO 'Champagne', but it could be registered for wine complying with the specifications of the PDO 'Champagne' (Class 33) and services in relation to wine bearing the PDO 'Champagne' (Classes 35 and 41). The Board noted that the protection of the PDO in question may be invoked not only against goods, but also against services[72] and that the PDO 'Champagne' enjoys an exceptional reputation in the EU and that there is a close link between the products covered by the PDO and the contested services. In Class 35, 'Nero Champagne' was liable to affect consumer preferences in respect of the *advertising; business management; business administration; office functions*, to which the image of a renowned product will inevitably be transferred and were not limited in relation to wine bearing the PDO Champagne. The use of the contested mark would benefit from the transfer of the distinctive qualities that the PDO 'Champagne' has acquired over the years, which would increase the ability of the contested mark to stand out from its competitors. It would also benefit from the transfer of the PDO image of a well-established category of prestigious wines with a long history in the EU market. Consumers would think that those services are specialised or specifically designed for companies wishing to do business related to or using Champagne wines. This constituted an exploitation of the reputation of the PDO 'Champagne'.

However, regarding the contested goods in Class 33 and services in Classes 35 and 41 that are limited to products that comply with the specifications of the PDO 'Champagne', the Board confirm that according to the limitation in the list of goods and services, use of the PDO is made in conformity with the product specification. By doing so, the Board confirmed the limitation theory of first instance.

Afterwards, *Champagne Nero* Decision was appealed before the General Court[73] by the *Comité interprofessionnel du vin de Champagne* and INAO who claimed its annulment, except to the extent the application for the trade mark at issue was rejected also for services in Class 35; the rejection the application for the trade mark at issue for the goods and services concerned in Classes 33, 35 and 41 or, alternatively, refer the case back to another Board of Appeal for reconsideration.[74]

[70] 19/09/2019, R 1223/2019-5, Mappa di Imola, para. 29.

[71] 17/02/2023, R 531/2022-2, NERO CHAMPAGNE / Champagne (GI).

[72] C-783/19 - Comité Interprofessionnel du Vin de Champagne, ECLI:EU:C:2021:713 para. 50–52.

[73] T-239/23, Comité interprofessionnel du vin de Champagne and INAO v EUIPO - Nero Lifestyle (NERO CHAMPAGNE).

[74] The information about the content of the appeal is available in curia website.

Because *Nero Champagne* was not yet final, a set of appeals were suspended in the Board of Appeal. Decisions on *Amarte Tequila 100% Agave (fig)*[75] *Adicción Tequila (fig),*[76] *Marqués de la Mancha*[77] were postponed until a judgment in case T-239/23 was delivered and become final. One of them, *Marqués de la Mancha*, has been already published and, following *Nero Champagne* judgment declared that the earlier PDO *La Mancha* was fully incorporated in the contested sign "Marques de la Mancha" with the result that there were visual and phonetic similarities and additionally the signs shared the same geographical name. The earlier goods wine was identical and similar to the contested goods wines and other alcoholic beverages (Class 33). Therefore the contested sign evoked the earlier PDO. The fact that the contested goods wines were limited to the compliance of the rules and conditions of the earlier PDO did not avoid the application of Article 103(2) of Regulation 1308/2013 .[78]

As regards the existence of evocation in cases of trade mark application with a limitation to wines produced according to the PDO rules, some national judgments can be referred at this stage, as example, in Spain[79] and France.[80]

---

[75] 07/08/2023, R 2248/2022-4, AMARTE TEQUILA 100% AGAVE (fig.) / TEQUILA (GI).

[76] 09/08/2023, R 2272/2022-4, ADICCION TEQUILA (fig.) / TEQUILA.

[77] 15/06/2023, R 1761/2022 - 5, MARQUÉS DE LA MANCHA / LA MANCHA DENOMINACION DE ORIGEN (fig.) et al.

[78] In *Marqués de la Mancha*, the Board of Appeal quoted that '54. However, the General Court recently clarified, in the judgment of 25/06/2025, 239/23-, NERO CHAMPAGNE/Champagne, EU:T:2025:638, § 49, that the situations listed in Article 103(2) of Regulation No 1308/2013 do not refer – with the exception of that provided for in Article 103(2)(a) (i) of that regulation – to comparable goods which do not comply with the specification. There is nothing in the wording of Article 103(2) (a) (ii) and (b) to (d) of Regulation No 1308/2013 to indicate that the circumstances set out therein cannot apply to the use of a PDO for products complying with the specification for that regulation'.

[79] The Spanish Supreme Court concluded that 'Lar de Duero' evokes the PDO Ribera del Duero even when the trade mark is applied for wines protected by the PDO Ribera del Duero. It's necessary for the Supreme Court that the trade mark applicant shows that they pertain to the producer organization PDO Ribera del Duero with an evidence of being registered in the Vineyards Registry or Cellars Registry (STS 1341/2021 - ECLI:ES:TS:2021:1341). Identical decisions were taken by the Supreme Court in the similar cases "CAVA BROT VINS DE TALLER" (TS 4359/2020 - ECLI:ES:TS:2020:4359); "CAVARQUIA BARCELONA" (STS 4116/2020 - ECLI:ES:TS:2020:4116); "DEEPSEACAVA" (STS 4121/2020 - ECLI:ES:TS:2020:4121) that evoked the PDO CAVA.

[80] The Judgment of 26 May 2023, Cour d'appel de Paris RG n° 21/09232, Pôle 5 - Chambre 2 which concludes that two trade marks with the term NewRhone, in any capacity whatsoever, in connection with wines, including those benefiting from the appellations 'Côtes du Rhône' or 'Côtes du Rhône Villages' evoke the PDO Côtes du Rhone and therefore have been cancelled. The term cannot be used as trade mark, commercial name, domaine name company name. A second pertinent decision to be referred is the one taken by INPI (26.8.2022, OP22-0433/NG) in relation with the figurative mark containing the term Cognapea, which evokes the PDO Cognac even for « Eaux de vie bénéficiant de l'indication géographique "Cognac" ».

Back to the substance of the *Nero Champagne* judgment[81] the Court noted, first, that EU law does not prohibit a trade mark from containing a PDO. However, its registration may be refused if the PDO does not comply with the product specification concerned, if its use exploits the reputation of a PDO or if the mark applied for conveys a false or misleading indication as to the provenance or origin of the product (para. 31). The General Court considered that there is a presumption that a trade mark containing a PDO and registered solely in respect of products complying with the specification of that PDO does not unfairly exploit the reputation of that PDO. However, such a presumption *iuris et de iure* may be rebutted where it is shown that a trade mark is liable to exploit unduly the reputation of a PDO, even if it designates only products which comply with the product specification of that PDO. Thus, when evidence to that effect is brought to EUIPO's attention, EUIPO must examine and assess whether there is an explotation of the reputation and consequently the resumption is rebutted(para. 54). Therefore, first, the Court finds that, despite EUIPO's consistent practice to that effect, the Board of Appeal erred in law in finding, in essence, that that presumption cannot be called into question by any evidence (para. 54). Secondly, if it is considered that the restriction is supported by a *iuris tantum* presumption, the Board of Appeal of EUIPO was in breach of its obligation to state reasons. It did not sufficiently explain how the evidence produced by the professional bodies was not capable of rebutting the presumption. EUIPO, and its Boards of Appeal, should have set out the reasons to a sufficient legal standard of why the reputation is not overturned (para. 66) due to the public interest that supports the recognition of the GIs.

Finally, the Court found that the word 'nero' could be perceived by consumers as evoking either the champagne grape variety or its colour, with the result that the mark applied for could convey false or misleading information. The Court notes that that term is used in the name of several well-known Italian grape varieties and that multiple grapevine varieties include it in their names. Furthermore, the word 'nero' will be understood by the Italian-speaking public as meaning 'black'. The public might therefore think that it is a 'black champagne', even though, under the specification of the PDO, a champagne can only be white or rosé (para. 81 to 85).

7.4.4 When Not to Look for a EUTM with GIs Reference: Descriptive and Non-distinctive

It is a well-established principle that a trade mark may consist of any sign capable of distinguishing the goods or services of one undertaking from those of other undertakings and that the possibilities are almost limitless. Generally, however, trader marks must not be of a such nature as to deceive the public as to, for instance, the nature, the quality or the geographical origin of the goods and the services.[82] It

[81] T-239/23 | Comité interprofessionnel du vin de Champagne et INAO v EUIPO – Nero Lifestyle (NERO CHAMPAGNE) EU:T:2025:638.

[82] O'Connors (2004), p. 110.

is not possible to register, as a trade mark, a term that merely consists of a geographical name as it would be unfair to others to give exclusive rights to one person.[83]

Any trade marks application when referring to a GI can potentially be examined under Article 7(1)(b) and 7(1)(c) EUTMR). In these cases, there is a risk of being considered descriptive and non-distinctive, like *Chateau de Pommard*.[84] On the one hand, the literal definition of the contested sign could and should lead to a refusal for registration pursuant to Article 7(1)(c) EUTMR, particularly as it follows therefrom, that the contested sign refers to any generic 'château' in the Pommard region, without any unique identification thereof. On the other, the applicant has not presented any arguments to address that specific finding.[85]

If the trade mark applied for consists solely of the geographical term, the application also falls under Article 7(1)(c) EUTMR, since it is considered descriptive of the geographical origin of the goods and of their quality. This means that the Office's objection will simultaneously raise absolute grounds for refusal under both Article 7(1)(c) and (j) EUTMR. According to *Chiemsee*,[86] Article 7(1)(c) EUTMR does not preclude the registration of geographical names which are unknown to the relevant class of persons—or at least unknown as the designation of a geographical location—or of names in respect of which, because of the type of place they designate, such persons are unlikely to believe that the category of goods or services concerned originates there'. For assessing the descriptiveness of geographical origin, the trade mark applied for must refer to a geographical location well-known by the general public. This public will do an automatic mental association between the geographical location and the pertinent category of goods and services, or this association can be done in the future. Thus, a mark consisting of a geographical term might be descriptive not only of the geographical origin of the goods and services but also of their quality or other characteristics.

For instance, as to *Grana Padano*,[87] the word mark Grana Padano (filed by Consorzio per la Tutela del Formaggio Grana Padano - Italian association for the protection of Grana Padano cheese) would be perceived as referring to the (main) subject matter of the services in question and not as an indication of commercial origin in the case of the wine or the cheese. Thus, it was partially refused for services in Classes 35, 41 and 43 and it was not filed as a collective mark.

---

[83] Gevers (1990), p. 285.

[84] 16/06/2020, R 2650/2019-5, Château de Pommard, para. 13.

[85] See Chapter 5 as regards the use of the term 'Chateau'.

[86] C-108/97 & C-109/97, Windsurfing Chiemsee, EU:C:1999:230.

[87] In 19/07/2019, R 2191/2018-2, Grana padano, the Board of Appeal concluded that the word mark GRANA PADANO (filed by Consorzio per la Tutela del Formaggio Grana Padano - Italian association for the protection of Grana Padano cheese will be perceived as referring to the (main) subject matter of the services in question and not as an indication of commercial origin in the case of the wine or the cheese. Thus, it was partially refused for services in Classes 35, 41 and 43 and it was not filed as a collective mark.

In another case of *Grana Padano* PGI, a figurative mark has been registered for the same Classes because of the existence of the figurative element.[88]

A search through past decisions on this topic reveals some interesting examples of trade marks containing the reference to the GI that were considered descriptive because they contained terms that were not identifying the commercial origin of the goods. Let's start with *Côtes du Rhône Vacqueyras*[89] where the refusal happened because the mere presence of the terms 'Côtes du Rhône' and 'Vacqueyras', which are both PDOs. We can follow with *Matprat*[90] where the EUTM application was refused because the Board of Appeal considered that the figurative trade mark contained the term PRAT, which is part of the protected designation Pollo y Capón del Prat. Finally, we find Café Gran Colombiano,[91] where the sign was rejected for all the contested goods in classes 30 (essentially various types of coffee and substitutes) and 32 (coffee bases/perfumed beverages).

To end up, according to EUIPO Guidelines, while restricting the relevant goods to a particular GI is usually a means of waiving the objection under Article 7(1)(j) EUTMR, it is irrelevant for Article 7(1)(c) EUTMR.[92]

7.5 Effective Advocacy Behind the Scenes: The Crucial Role of GIs' Protection in Inter Partes Proceedings

Protection of GIs on relative grounds is complementary to the *ex officio* protection on absolute grounds (Article 7(1)(j) EUTMR). While there is interaction and overlaps with protection under absolute grounds, there are differences both in terms of procedure and substance arising from the relative grounds nature of protection under Article 8(6) EUTMR.[93]

In *inter partes* proceedings at EUIPO, the GIs producers have the possibility of opposing any EUTM application under the conditions and requirements fixed in

[88] In 15/11/2023, R 1073/2022-5, GRANA PADANO (fig.), the collective mark applied for is not identical to the PDO 'Grana Padano' and it is used in a different way than the relevant PDO, as described by the regulation of use, read in the context of the PDO's product specification. Therefore, it does not deceive the public regarding the collective nature of the mark. For the same reasons, and considering its figurative element, it does not lack of distinctive character (Grana Padano).

[89] 21/02/2018, R 2305/2017-5, CRU CÔTES DU RHÔNE VACQUEYRAS (fig.), para 45.

[90] 23/03/2018, R 26/2018-5, MATPRAT (fig.), para 28.

[91] 07/12/2018, R 477/2017-1, CAFE Gran Colombiano con todo el aroma y sabor del buen café (fig.), para. 42.

[92] For example, an application for the word mark 'Bergerac' for wines will simultaneously be objected to under both Article 7(1)(c) and (j) EUTMR: it consists solely of the PDO 'Bergerac' and is therefore descriptive. If the goods are subsequently limited to 'Bergerac' (GI) wine, the objection under Article 7(1)(j) EUTMR will be waived but the trade mark will still be descriptive and can be objected to under Article 7(1)(c) EUTMR. See EUIPO Guidelines Part B Examination, Section 4 Absolute Grounds of Refusal, Chapter 10.4.1. Use of a GI (direct or indirect use).

[93] See EUIPO Guidelines Part C Opposition, Section 6 Geographical Indications (Article 8(6) EUTMR) para. 1.

Article 8(6) EUTMR.[94] In this context, it is relevant to determine whether the EU Law conferred on the proprietor of the PDO/PGI the undisputed right to prohibit the use of the later mark and, consequently, to assess, on the basis of that EU Law, whether the PDO/GI could oppose the trade mark applied for.[95]

Pursuant to Article 8(6) EUTMR, upon opposition by any person authorised under the applicable law to exercise the rights deriving from a GI, the trade mark applied for shall not be registered where and to the extent that, pursuant to Union legislation or national law providing for the protection of GIs:

- an application for a PDO or PGI has already been filed in accordance with Union legislation or national law prior to the date of application for registration of the EU trade mark or the date of the priority claimed for the application, provided that it is subsequently registered,
- the PDO or PGI concerned confers the right to prohibit the use of a subsequent trade mark.

Consequently, for this provision to apply, the following conditions must be met:

- The opponent must be authorised under the relevant law to exercise the rights arising from a PDO or a PGI;
- The application for the PDO or a PGI was submitted, in accordance with Union legislation or national law, prior to the date of application for registration of the EU trade mark or the date of the priority claimed for the application, subject to its subsequent registration;
- In accordance with Union legislation or national law, that PDO or PGI confers the right to prohibit the use of a subsequent trade mark[96] in situations covered by Article 26 Regulation (EU) 2024/1143 or situations protected under national law or international agreements, including agreements concluded by the EU with third countries. An opposition based on Article 8(6) EUTMR can be successful against goods and services that are not comparable to those for which the GI is protected, provided that the opponent submits evidence and arguments to prove that either Article 11(1)(a)(ii) or Article 11(1)(b) of the Geneva Act apply to such

---

[94] Regulation (EU) 2015/2424 amending Council Regulation (EC) No 207/2009 on the Community trade mark (Amending Regulation) introduced Article 8(4a) of Regulation No 207/2009 [now Article 8(6) EUTMR] as a specific ground of opposition for GIs. GIs can no longer form the basis of an opposition under Article 8(4) EUTMR, even though the wording of that provision has not changed. However, if an opponent files an opposition based on a GI after entry into force of the Amending Regulation, in which it indicates incorrectly Article 8(4) EUTMR as a ground for opposition, the Office will examine the opposition, to the extent that it is clearly based on a GI, as if the ground invoked were Article 8(6) EUTMR. In such a case, the notice of opposition leaves no doubt as to the opponent's intention to invoke the ground for opposition protecting earlier GIs See EUIPO Guidelines 2020 https://guidelines.euipo.europa.eu/1803468/1785456/trade-mark-guidelines/5-1-relationship-between-article-8-4%2D%2Dand-8-6%2D%2Deutmr.

[95] 18/09/2015, Federación Nacional de Cafeteros de Colombia/OHIM — Accelerate (COLOMBIANO COFFEE HOUSE), T-359/14, not published, EU:T:2015:651, para. 26.

[96] 29/08/2022, R 462/2022-5, GRILLOUMAKI / Χαλλουμι' (Halloumi)/'Hellim para. 15.

> non-comparable goods or services.[97] The Office will therefore apply the same standard to GIs protected under the Geneva Act and will examine potential conflict with earlier GIs, bearing in mind, mutatis mutandis, the rules contained above in the Guidelines.[98]

Thus, it can be asserted that an opposition against an EUTM application that has been appropriately limited cannot be sistematically rejected. In July 2026, a new version of the EUIPO Guidelines facilitates the substantation of EU GIs in Article 8(6) EUTMR as EUIPO has *ex officio* knowledge of the Union Registers. It determines that the presumption is rebutted if it is shown that the trade mark is likely to 'unduly exploit the reputation of a GI' on the basis of concrete, substantiated and consistent information presented before the EUIPO.[99]

7.5.1 Direct and Indirect Use of a GI and the Difficulty of Demonstrating Likelihood of Confusion: *Campo Verde, Toro de Piedra, Colombueno, Chenine*, etc.

Under Article 26 (1) of Regulation (EU) No 2024/1143, GIs are protected against any direct or indirect commercial use of the GI in respect of products not covered by the registration, where those products are comparable to the products registered under that name or where use of that geographical indication for any product or any service exploits, weakens, dilutes, or is detrimental to the reputation of, the protected name, including when those products are used as an ingredient.

The direct or indirect use happens when there is a high degree of visual and aural similarity and the products are identical or there is a commercially relevant ingredient, or they are comparable or the similarity is not as regards the product but related any services which may be associated with the direct or indirect distribution of those products. As a consistent solution to the *inter partes* conflict, when there is a discussion on the possible direct or indirect use of a GI, the Board of Appeal conclude that there is likelihood of confusion on the basis of the goods being identical

[97] See EUIPO Guidelines Part C, Opposition, Section 6, Geographical Indications, paragraph 3.2, Scope of protection of GIs protected under national law or international agreements. The Office considers that the notion of 'goods that are not of the same kind' is analogous to the notion of 'non-comparable goods' under the EU Regulations providing for the protection of GIs,

[98] See EUIPO Guidelines, Part B, Examination, Section 4, Absolute grounds for refusal, Chapter 10, Trade marks in conflict with geographical indications (Article 7(1)(j) EUTMR), para. 4.

[99] See EUIPO Guidelines, Part C, Opposition, Section 6, Geographical Indications, paragraphs 2.4 Specifics of substantiation and 3.1.3. Limits to the scope of protection of GIs on relative grounds.

(*Organic Quijote*[100]) or comparable (*Sol de Mallorca*[101]) even though the EUTM application contains a limitation (*Fratelli Asti*[102]).

In *By Latúe PRO & ECO* (fig.)[103] the opposition was based on the PDO Prosecco. The EUTM application was considered visually highly similar and sounded almost identical to Prosecco, in Romanian and in Hungarian, and the Board of Appeal decided that it would be perceived as an intelligent misspelling of the PDO Prosecco, thus amounted to 'direct use', which was likely to exploit the repute of the PDO. The Board stated that the services applied for in Class 40, even after their limitation' were expressly related to 'wines' and cover all the spectrum of activities of a wine-maker, from the pressing of grapes, to their transformation or processing to make and age wine (and related alcoholic beverages). All the services for which the EUTM was applied for were considered in the realm of the typical activities of 'Prosecco' wine producers who can legitimately use the PDO 'Prosecco', only if they obtain, produce, and process their wine in accordance with the strict requirements laid down in the specifications to produce quality wines with the PDO 'Prosecco'.

However, the absence of likelihood of confusion, and the subsequent rejection of the appeal filed by the PDO, was concluded in several cases as in *Campo Verde*[104] or *Green Choice* (fig)[105] in relation to the PDO Vinho Verde, because, being the goods identical, the relevant consumer will not believe that the goods come from the same undertaking or from economically linked undertakings.

In Toro de piedra,[106] the Board of Appeal stated, firstly, that the appellant (PDO Toro) was not able to proof genuine use of its earlier figurative marks, as the two trade marks were not used with respect to the goods protected in accordance with their essential function, the indication of origin; secondly that, since the compared signs were visually, aurally and conceptually dissimilar, no likelihood of confusion could arise upon Article 8(1)(b) EUTMR; and, thirdly that the appellant did not

[100] 04/04/2017, R 1428/2016-2, ORGANIC Quijote (fig.) / LA MANCHA DENOMINACIÓN DE ORIGEN (fig.) et al. para. 44.

[101] The Board of Appeal considered that 'Alcohol free wine' is a comparable product to 'wine' because falls under the broader category of 'soft drinks'. Since the sign applied for contains a protected geographical indication, there is a direct use of the PDO for wines 'Mallorca', the conditions of said article are fulfilled and the opposition must be allowed under Article 8(6) EUTMR (09/02/2021, R 237/2020-4, Sol de Mallorca / Mallorca) para. 35.

[102] 21/06/2018, R 2079/2017-1, CASA FONDATA NEL 1991 FRATELLI VINO BIANCO FRATELLI Asti (fig.) / CONSORZIO DELL'ASTI D.O.C.G. (fig.) et al., para. 33.

[103] 14/11/2019, R 425/2019-1, By Latúe PRO & ECO (fig.) / PROSECCO PDO PROSECCO DOC (fig.) et al, para 49.

[104] 02/10/2020, R 282/2020-5, Campo Verde / Vinho verde et al. para 78. The Board of Appeal decided to endorse the rejection of the opposition as unfounded on the basis of Article 8(6) EUTMR for the following reasons: a) as regards the PDO Vinho Verde the opponent ought to have produced evidence that it was entitled to file the opposition on the basis of such earlier right and such evidence was not submitted.

[105] 07/04/2021, R 1661/2020-5, GREEN CHOICE (fig.) / Vinho verde et al, para. 71.

[106] 03/09/2020, R 2751/2019-4, TORO DE PIEDRA (fig.) / Toro denominación de origen et. Al., para 51.

submit any evidence of enhanced distinctive character or of reputation of its earlier trade marks, so Article 8(5) EUTMR was also not applicable. Therefore, the Board dismissed the appeal.

In *Delicius Green Eva Fricke,*[107] PDO Vinho Verde relied erroneously on the national registrations of the GI for wines, which since the uniform protection exist in the EU cannot invoked as basis of the opposition under Article 8(6) EUTMR. The Board of Appeal concluded that there was not likelihood of confusion under 8(1)(b) EUTMR even for identical goods, since the signs were clearly visually and aurally dissimilar, while there was only low conceptual similarity if any.

Moreover, in *Portsoy*[108] the contested 'Scotch Whisky complying with the specifications of the PGI *Scotch Whisky* was considered not comparable to 'wine' protected by the PDO 'Port'. The Board of Appeal considered that the mere fact that the earlier PDO 'Port' and the contested sign 'Portsoy' coincide in four letters ('Port') was not sufficient for the relevant public to make a clear and direct connection between the contested sign and the goods that were not comparable with 'wine' of the earlier PDO 'Port'. That assessment was confirmed by the not insignificant differences between 'wine' and 'Scotch Whisky complying with the specifications of the PGI Scotch Whisky' in terms of, inter alia, ingredients, alcohol content, taste, methods of production and use. As said before, the Court confirmed the decision adopted by the Board of Appeal.[109]

In *Colombueno,*[110] the Board of Appeal quoted case-law that provided clarification with respect to the concepts of direct and indirect commercial use,[111] and reached as conclusion that the contested mark, by making direct use of the reputed name of the GI product (and already by this very use triggers the image of the protected product in the consumer's mind), is likely to affect the consumer's preferences with respect to the services applied for (Class 43 - Services for providing food and drink; temporary accommodation) to which the image of a product that is not only reputed, but is considered a symbol of Colombia, will inevitably be transferred.

As quoted in previously, the Board of Appeal recalled the EU legislator stated that GIs can be registered because of their reputation. This reputation was argued by

---

[107] 18/02/2021, R 1720/2020-5, Delicious Green Eva Fricke (fig.) / Vinho verde et al., para. 64.

[108] 14/11/2023, R 1885/2022-4, PORTSOY / Port., para. 35.

[109] T-40/24, IVDP v EUIPO - The Benriach Distillery Company (PORTSOY), para. 73.

[110] 01/10/2021, R 566/2015-2, COLOMBUENO / CAFÉ DE COLOMBIA, para. 67.

[111] C-44/17, Scotch Whisky Association, ECLI:EU:C:2018:415, para. 39. In Scotch Whisky Association, the Court has clarified that a distinction must be drawn between situations in which the use is 'direct' and those in which it is 'indirect'. More specifically, there is 'indirect commercial use' of a Gi, when the disputed element is used in a form that is 'either identical to that indication or phonetically and/or visually similar to it' This is particularly clearly stated in the Advocate General's opinion in Scotch Whisky: "unlike 'direct' use, which implies that the GI is affixed directly to the products in question or their packaging, 'indirect' use requires the indication to feature in supplementary marketing or information sources, such as an advertisement or documents relating to it" AG Opinion Case C-44/77 Scotch Whisky Association ECLI:EU:C:2018:111, para. 30.

the producer groups when applying for a recognition of the GI.[112] Therefore, GIs are intrinsically reputedmerely because they are registered. This is independent of whether a GI has been registered based on a claim of reputation essentially attributable to its geographical origin.

When dealing with the protection, there is a second and different idea of reputation, when it is said that the GI is protected against undue exploitation of their reputation (Article 26 Regulation (EU) 2024/1143). In this second scenario, with respect to the assessment of reputation under Article 8(5), the Board of Appeal referred to apply a trade mark method. This method is referred in Article 8(5) EUTMR which provides as follows:

> 5. Upon opposition by the proprietor of a registered earlier trade mark within the meaning of paragraph 2, the trade mark applied for shall not be registered where it is identical with, or similar to, an earlier trade mark, irrespective of whether the goods or services for which it is applied are identical with, similar to or not similar to those for which the earlier trade mark is registered, where, in the case of an earlier EU trade mark, the trade mark has a reputation in the Union or, in the case of an earlier national trade mark, the trade mark has a reputation in the Member State concerned, and where the use without due cause of the trade mark applied for would take unfair advantage of, or be detrimental to, the distinctive character or the repute of the earlier trade mark.

As regards the trade mark method, the Board of Appeal recalled in *Chenine*[113] that Article 8(5) EUTMR only apply where the following conditions are met:

- The signs must be identical or similar.
- The opponent's trade mark must be well known. That reputation must pre -date the filing date of the contested trade mark, must exist in the relevant territory and must relate to the goods and/or service on which the opposition is based.
- Risk of detriment: the use of the contested trade mark must take unfair advantage of, or be detrimental to, the distinctive character or the repute of the earlier trade mark.

As stated by the case law, the aforesaid conditions are cumulative and failure to satisfy any of them means that the opposition must be rejected pursuant to Article 8(5) EUTMR.[114]

However, it should be noted that compliance with these requirements may not be sufficient. Thus, the opposition may fail if the applicant successfully demonstrates due cause for use of the contested trade mark.

Therefore, the complicated issue here is to furnish the required evidence to proof reputation and the evidence to show that use of the contested trade mark would take unfair advantage of or be detrimental to the distinctive character or reputation of the GI. For instance, in *Chenine*, the Board of Appeal considered that PDO Cava failed

---

[112] Article 9 Regulation (EU) 2024/1143.

[113] 22/04/2022, R 981/2021-1, CHENINE Brut PREMIUM CAVA BOTTLE (3D)/Council REGULADOR CAVA (fig.) et al., para. 40.

[114] 16/12/2010, T-345/08, & T-357/08, Botolist/Botocyl, EU:T:2010:529, para. 41.

to proof both requirements; first, in the notice of opposition and in the notice of appeal, the opponent alluded to the suggestive nature of the use of the term 'CAVA' in the contested trade mark but this claim was not accompanied by further arguments unless it is connected with its deceptive nature. Second, the contested trade mark related to goods that comply with the PDO Cava and the opponent had not demonstrated that any infringement proceedings had been initiated due to the fact that the wine marketed did not comply with the specifications and the specification of goods of the PDO Cava, at the time of filing of the EUTM application or at the time of the opposition. By virtue, thereof, evocation was not applicable. Similar conclusions were reached in the following Decisions: *Beher Guijuelo* (where no proof of reputation was submitted so Article 8(5) was neither in discussion),[115] *Campo verde*[116] and *Toro de piedra*.[117]

In *Chenine*, the Board of Appeal remarked that the argument of exploitation of the reputation of the PDO can only be invoked if the use of the PDO for products not covered by the right is demonstrated. In the case at hand, the goods designated by the contested trade mark comply with the specifications of the PDO Cava, thus fulfilling the specific function of PDOs, consisting of designating the geographical origin and particular qualities of the goods. Therefore, in the case at hand, the Board of Appeal concluded that there was no exploitation of the reputation of the PDO Cava and the decision was not appealed and become final.

7.5.2 When a Misuse Could Happen: *Colombiano Coffe House*

As per Article 26 (1) of Regulation (EU) No 2024/1143, GIs are protected against any misuse, imitation or evocation, even if the true origin of the products or services is indicated or if the protected name is translated, transcribed or transliterated or accompanied by an expression such as 'style', 'type', 'method', 'as produced in', 'imitation', 'flavour', 'like' or similar, including when those products are used as an ingredient.

In *Colombiano Coffe House*,[118] the Board of Appeal set up that misuse of the GI could happen when using the GI in an inappropriate context, which would be detrimental to the reputation of quality products (e.g. Café de Colombia, for industrial paintings). The Board of Appeal applied the BNIC Cognac[119] criteria and found that that the goods in Class 30, apart from tea and cocoa, could not be comparable to coffee. A fortiori, for the Board of Appeal, the contested sign's services in Class 43 could not be comparable products since their link with coffee was vague and imprecise. As for tea and cocoa, which the Board of Appeal deemed not to be comparable

[115] 03/07/2019, R 942/2018-4, BEHER Bernardo Hernandez BEHER GUIJUELO (fig.) / Guijuelo Denominación de Origen (fig.) et al.

[116] Op. cit. 102.

[117] Op. cit. 104.

[118] 30/11/2018, R 251/2016-1, COLOMBIANO COFFEE HOUSE (fig.). para. 132.

[119] C-4/10 - Bureau National Interprofessionnel du Cognac, ECLI:EU:C:2011:484, para. 54.

to coffee, they had certain elements in common with coffee, such as their physical appearance as well as their method of use (all can be in soluble powder form to be mixed with boiling water in order to make a hot drink).

Nevertheless, for the Board of Appeal, these common features would not justify the inclusion of these products in the same class for the purposes of protection of a GI. This reference to "the same class for the purposes of protection of a PGI" is problematic, as the term class is not defined under GI law, is a term of art in trade mark law which clearly does not apply in the assessment of comparability, and seems to indirectly introduce a new, potentially relevant criteria for the assessment of comparability. In the absence of further case law relying on this criterion, this statement should be taken as *obiter dicta*, in the light of the Board's subsequent reasoning based on the *BNIC* case criteria: whether they are frequently distributed through the same channels and whether they subject to similar marketing rules.

7.5.3 GIs' Imitation or Evocation: From *Romeral de Jabugo* to *Padrón Auténtico and Marqués de la Mancha*

As quoted before, Article 26 (1) (b) Regulation (EU) 2024/1143 states that GIs cannot be imitated or evoked, even if the true origin of the products or services is indicated or if the protected name is translated, transcribed or transliterated or accompanied by an expression such as 'style', 'type', 'method', 'as produced in', 'imitation', 'flavour', 'like' or similar, including when those products are used as an ingredient.

An undisputed Board of Appeal Decision that sheds light on the consideration to be given to the elements making up a trade mark in order to determine whether there is a GIs evocation is *Romeral de Jabugo*.[120] In this case, the Board of Appeal identified that there was a close connection between the contested services and the goods for which the PDO is protected, and since "Jabugo" is a famous PDO in Spain; the Spanish public may think of it in relation to the contested services, so the use of the contested mark would exploit the reputation of the protected name.

For the Board of Appeal, it was important to highlight that this was demonstrated, among other, in the sales figures and the economic value of the production of ham and pallets under the PDO; in the distinctions, prizes and awards at regional, national and international level; in the press articles and promotion of PDOs at national level; and in the press articles for PDOs at national level. Further, the decision determines that extra steps can be required when assessing evocation. Aligned to the case-law,[121] the Board of Appeal interpreted that there might be evocation or imitation of a PDO/PGI where:

[120] 20/04/2022, R 1296/2021-4, Romeral de jabugo / Jabugo, para. 52.

[121] C-132/05, Commission v Germany, EU:C:2008:117, para. 44; C-87/97, Cambozola, EU:C:1999:115, para. 25 C-44/17, Scotch Whisky Association, EU:C:2018:415, para. 44-45; and C-75/15, Viiniverla, EU:C:2016:35, para. 21 and the case -law cited therein.

- the EUTM incorporates the significant part of the PDO/PGI (the part of the PDO/PGI from which the consumer identifies the PDO/PGI and which is not the generic element);
- the EUTM contains an equivalent adjective/noun that indicates the same geographical origin.
- the PDO/PGI is translated.
- the EUTM includes a 'de-localiser' expression in addition to the PDO/PGI or its evocation.

Consequently, in the referred *Romeral de Jabugo*, the Board of Appeal pointed out that the concept of evocation provided for by the relevant legislation also covers cases where the term is used to designate a product that incorporates part of a protected name, so that the consumer, when faced with the name of the product, is led to have in mind, as a reference image, the product covered by the name. Furthermore, according to case -law, the concept of 'evocation' may exist even in the absence of such aural and visual similarities between the signs. In fact, in addition to the above criteria, account must be taken, where appropriate, of the criterion of 'conceptual proximity'. Therefore, this decision arguably provides legal certainty for future trademark owners.

In *Puerto fino*,[122] the Board of Appeal found that the mark Puerto fino, consisting of the word 'puerto' (the Spanish equivalent of 'porto') and the word 'fino' (a traditional term used for centuries in conjunction with Porto/Port wine) would evoke the earlier GI 'Porto'/'Port', at least for the Portuguese-speaking consumer.

In *Piemontino*,[123] the application was rejected because it contained a reference to the Italian region 'Piemonte' and thus was evoking the PDO 'toma piemontese' protected for cheese.

In *Cave de Tain*,[124] the General Court confirmed the Board of Appeal Decision about the absence of likelihood of confusion with the GI Cava, because GI's protection '*does not extend against evocation when the differences between the signs, like figurative elements or conceptual meaning, precluded evocation. The visual and phonetic differences between the signs at issue were such as to preclude, in any event, the possibility of evocation*'. A decision in the same vein can be found in *Las Villas*.[125]

The concept of 'evocation' covers a case in which the term used to designate a product incorporates part of a GI, so that it leads the consumer, upon seeing the name of the product, to think, as a reference image, of the product which benefits from that GI.

In *Padrón Auténtico*,[126] the Board of Appeal considered that the mark Padrón Auténtico was evoking the PDO 'Pemento de Herbón' because a significant part of

122 07/03/2019, R 831/2018-2, PUERTO FINO (fig.), (PUERTO FINO / PORTO). Para. 28–31.

123 16/07/2018, R 2110/2017-1, PIEMONTINO (fig.) para. 38.

124 T-774/16, Cave de Tain, ECLI:EU:T:2018:441, para. 44–52.

125 01/09/2021, R 1734/2020-2, Las villas, para. 27–28.

126 16/07/2021, R 879/2020-5, Padron AUTÉNTICO (fig.) / Pemento de Herbón., para. 38.

the relevant consumers, including significant part of the Spanish public would perceive the sign at hand as 'the genuine peppers of Padrón', or in other words, that these were authentic Padrón peppers and quoted *Camboloza* and *Parmesan* case law.[127]

The Board remarked that this connotation may also encompass the geographical origin of the peppers in question, namely 'Padrón', a municipality in the province of La Coruña in Spain. This interpretation was consistent with the case law on the concept of 'evocation' that covers a situation in which the term used to designate a product incorporates part of a protected designation, so that on seeing the name of the product, the consumer will think, as a reference image, of the product which benefits from the designation.

Likewise, the approach to the concept of evocation has been consistently applied and reaffirmed in various notable decisions, including, but not limited to, the following:

- *Valdobbiadene*,[128] where, considering the already claimed and irrefutable reputation of the PDO 'Valdobbiadene Prosecco', the similarity between the PDO and the contested sign, and the identity and comparable nature of the goods in question, the Board of Appeal considered that, for a significant proportion of the European public, the public with a reputation for PDOs, there will certainly be instances of evocation. In fact, for the Board of Appeal, this part of the European public would have perceived the contested trade mark as a neologism between the term 'Valdobbiadene' and the term 'Prosecco'. The public would therefore be able to believe that the purpose of the sign 'VALSECCO'—regardless of the actual intentions of the applicant—was to evoke the opponent's PDO. The circumstance alleged by the applicant that the word 'secco' means 'dry' did not preclude, in view of the Board of Appeal, such evocation in the mind of the public that includes all consumers in the EU and not only the Italian-speaking ones. For these reasons, the Board considered that the sign 'VALSECCO' was a clear evocation of the PDO.
- *Es Valentía*,[129] where the Board of Appeal decided that the contested sign would have evoked the PDO 'Valencia' in the minds of part of the Spanish-speaking and non-Spanish-speaking consumers in the EU. The use of the word 'Valentía' will be perceived as a clever emphasis on the bravery of the inhabitants, which is the origin of the ancient Roman name of the city of Valencia. This did not exclude but reinforced the evocation of the PDO 'Valencia'.
- *Condiparma*.[130] In the case at hand, the Board of Appeal concluded that Italian-speaking consumers will identify the term "PARMA" in the contested sign's

[127] C-87/97, Cambozola, EU:C:1999:115, para. 25; C-132/05, Commission v Germany, EU:C:2008:117, para. 44.

[128] 07/06/2021, R 1249/2016-1, VALSECCO / VALDOBBIADENE PROSECCO et al., para. 28.

[129] 06/04/2021, R 885/2020-1, Es Valentía / VALENCIA DENOMINACION DE ORIGEN (fig.) et al., para. 63.

[130] 07/11/2023, R 937/2022-4, CONDIPARMA (fig.) / PARMIGIANO REGGIANO (GI), para. 70.

verbal element "CONDIPARMA", the name of the well-known Italian city and province. A non-negligible part of average Italian consumers will perceive the term "CONDI" in the contested sign as referring to the verb "condire" (to season). The contested sign was visually and phonetically similar to a low degree and conceptually similar to a medium degree to the PDO on account of the common reference to "Parma". Since the contested "milk products" contained the goods protected by the PDO, namely, cheeses, for those goods, the contested sign would evoke in the relevant public a sufficiently clear and direct link with the PDO relied on, which has a high reputation for cheeses. However, with reference to the contested good "milk", the Italian public would recognise the name of the Italian city/province PARMA in the contested sign as a mere reference to the geographical origin of that product, and not as an evocation of the PDO "Parmigiano Reggiano". The contested sign was therefore rejected for "milk products"

- *Cavca.*[131] In this case, the Board concluded that there was some conceptual proximity between the terms firstly because the relevant public may think that the contested mark is a misspelling of the CAUCA verbal element of the GI and, therefore, the existence of evocation of the earlier PGI and secondly because the products and services of the contested mark were comparable to coffee.[132]
- *Morellone le caniette (fig).*[133] The Board of Appeal decided that or the EU public outside Italy the verbal element 'MORELLONE' was similar to the component in PDO 'MORELLINO'. This component was not negligible and at least for those consumers was not meaningful. The specific circumstances of the present case led to conclude that the opposition pursuant to Article 8(6) EUTMR was well founded.
- *The grand wines GW Rioja Alavesa (fig).*[134] In the Board' opinion, the contested trade mark, by using the reputed name of the PDO 'RIOJA' (and by that same use already activates the image of the protected product in the mind of the relevant public), was likely to affect the preferences of the consumer with respect to the contested services, to which the image of a well-known product would inevitably be conveyed.
- *Macroon Bufalloumi.*[135] Concerning the evocation, the Board of Appeal stated that the contested mark reproduces partly the PDO, was visually and aurally

[131] 13/10/2023, R 825/2023-5, CAVCA / Café de Cauca (GI), para. 26.

[132] In this case, the application was submitted for the following goods and services: Class 11: Electric coffee brewers; Coffee roasters; Expresso coffee machines; Coffee capsules, empty, for electric coffee machines; Domestic coffee percolators [electric]; Electric coffee filters; Class 35: Rental of card-operated vending machines; Class 43: Café services; Services for providing drink; Serving drink for guests; Serving drinks; Providing drink services; Catering in fast-food cafeterias.

[133] 26/11/2022, R 2242/2020-1, MORELLONE LE CANIETTE (fig.) / Morellino et al., para. 34.

[134] 01/09/2022, R 1714/2021-5, THE GRAND WINES GW RIOJA ALAVESA (fig.) / Rioja (DENOMINACION DE ORIGEN PROTEGIDA) et al., para. 63.

[135] 29/07/2022, R 1940/2021-5, Macroom buffaloumi / Halloum, para. 63.

similar to it and having regard to the fact that the contested goods were foodstuff and agricultural goods there was an evocation of the earlier PDO.
- *Halime (fig).*[136] In the view of the Board of Appeal, the contested mark reproduced partly the PDO Halloumi, was visually and aurally similar to it and having regard to the fact that the contested goods were foodstuff and agricultural goods there was an evocation of the earlier PDO.
- *Parmatutto*[137] where the Board pointed out that the protection clearly extends to products which were comparable to the products registered under that name or where using the name exploits the reputation of the protected name, including when those products were used as an ingredient.
- In the case of *Bolgaré (fig)*,[138] the sign applied for contained the word *Bolgaré*, which—contrary to the findings of the contested decision—the Board considered that it constituted an evocation of the PDO *Bolgheri* for wine from the perspective of the Italian public. The signs were highly similar, both visually and phonetically. In view of the Board of Appeal, contrary to the applicant's arguments, the relevant Italian consumer would not associate *Bolgaré* with Bulgaria, also because Bolgaré appears to be an Italian word. This was applied to all of the contested goods, since they were all either identical to wine, contain wine or were in proximity to wine.

These decisions of the Boards of Appeal, among others, have reproduced and reinforced the approach to the concept of evocation, providing a consistent and predictable framework for the assessment of evocation in various contexts.

A decision in the same vein was taken in *Perisecco*[139] as regards PDO *Prosecco* (wine) where, bearing in mind all the relevant factors of the case at hand, the Board of Appeal concluded that there was a strong likelihood that a substantial part of European consumers would see the contested mark 'PERISECCO' as an intelligent misspelling of the PDO *PROSECCO*, thus the contested mark will at least evoke in the mind of the average European consumer throughout the EU, the PDO 'PROSECCO'. Same conclusion on evocation was adopted in PriSecco[140]

Likewise, in *Queso y Torta de la Serena*[141] as to PDO *Torta del Casar* (cheese). The Board of Appeal pointed out that, where the consumer evokes it might think, as a reference image, of the product that benefits from the PDO, and not at its place of manufacture, that there might be evocation even if the consumer perceives that the

[136] 25/07/2022, R 1939/2021-5, Halime (fig.) / 'ΧΑΛΛΟΥΜΙ' (HALLOUMI)/'HELLIM', para. 68.

[137] 20/01/2014 – R 1900/2013-5 – PARMATUTTO, para. 25.

[138] 21/03/2022, R 2564/2019-2, BOLGARÉ (fig.) / Bolgheri et al., para. 31.

[139] 05/11/2021, R 1103/2019-1, PERISECCO (fig.) / PROSECCO (other) et al. para. 80. Bearing in mind all the relevant factors of the case at hand, there is a strong likelihood that a substantial part of European consumers would see the contested mark 'PERISECCO' as an intelligent misspelling of the PDO PROSECCO, thus the contested mark will at least evoke in the mind of the average European consumer throughout the EU, the PDO 'PROSECCO'.

[140] 30/05/2024, R 1454/2022-5, PriSecco / 'Prosecco', para. 93–94.

[141] 25/02/2021, R 696/2018-5, QUESO Y TORTA DE LA SERENA (fig.) / Torta del Casar et al, para. 116.

expressions *in discordia* refer to different geographical locations and that evocation is in no way limited to cases where the element evoked is geographical in nature. In this case, the Board followed the reasoning established by the Court[142] to conclude that the term "Torta" (in relation to "Quesos") was not or has not become generic in nature, and that, in addition, taking into account the evidence provided by the opponent plus the expert report requested ex officio, the term "Torta" of the GI "Torta del Casar" is a traditional non-geographical name, also covered by GI protection. The Board also confirmed that the requirements of Article 8(6) (Article 8(4) then) were satisfied and, thus, the trademark application fall under the relative ground for refusal of the insofar as its use would constitute an "evocation" of the GI "Torta del Casar".

It's part and parcel that evocation happens although the goods are not comparable at all; for instance, in *Champagnola*[143] and *Dutch Genquila.*[144] As for 'Champagnola', it was concluded that it clearly evoked 'Champagne' even though contested goods and services (bakery goods and services) were not 'comparable' to wine. Regarding *Dutch Genquila*, the point of departure of the analysis was the visual and phonetic proximity between the PGI 'Tequila' and the contested mark 'Dutch Genquila' which reproduced the characteristic ending 'quila' of the PGI 'Tequila'. For the Board, even if 'wine' and 'spirit drinks' were not comparable and the legal provisions on evocation might be interpreted in the sense that there was no prerequisite that the product covered by the protected designation and the goods or services covered by the contested sign were 'comparable' or 'similar' to establish evocation. In this sense, the Board of Appeal considered that when the relevant EU consumer of the goods at issue, who is reasonably well informed and reasonably observant and circumspect, was confronted with the sign 'Dutch Genquila', the image triggered directly in his/her mind is that of the product whose GI is protected, namely 'Tequila' spirit drinks. Therefore, from the perspective of the EU public, the contested sign did constitute an evocation of the earlier PGI Tequila.

Nevertheless, the mere similarity between the marks alone cannot justify any imitation or evocation of the earlier right. For instance, in *Halloumaki* (fig)[145] the earlier goods, 'cheese' which are consumed by the general public, were completely different to the contested 'business services' which target a professional public. The opponent was under the obligation to substantiate its position with more specific arguments and reasoning with regard to the confronted goods and services, and to explain how that alleged imitation or evocation would occur. In the absence of any such explanation, considered the Board of Appeal, the opposition under Article 8(6) EUTMR must have failed.

In this context, sometimes the inter partes proceedings have had different outcomes as regards the same GI because of different factors; for instance, *Porte*

[142] T-828/16, Consejo Regulador de la Denominación de Origen "Torta del Casar" v European Union Intellectual Property Office, ECLI:EU:T:2017:918.

[143] 17/04/2020, R 1132/2019-4, Champagnola / Champagne, para. 44.

[144] 06/03/2024, R 1033/2023-2, DUTCH GENQUILA / TEQUILA (GI), para 20.

[145] 12/05/2022, R 1941/2021-5, HALLOUMAKI (fig.) / Halloumi, para. 29.

Noire,[146] *Portwo Gin*[147] or *Port* ruighe[148] where the Board considered that the use of the contested sign containing the element 'Port' in relation to spirits was likely to take undue advantage of, and therefore exploit, the exceptional repute enjoyed by the wines with PDO 'Porto', among European consumers.

In *Portwo Gin,* the General Court endorsed the decision by confirming that the Board of Appeal was correct to note that, visually, the first four letters of the mark applied for were identical to the first four letters of the PDO 'Porto' and that the initial part of a sign was generally what caught the attention of the consumer. Moreover, it would not rule out that a significant part of the relevant public might perceive the presence of the letter 'w' as a spelling mistake. As to the phonetic aspect, in the view of the General Court, the Board of Appeal was right to consider that, at least in French, the sound of the initial element of the mark applied for was identical to the sound of the PDO 'Porto'. As regards the additional word 'GIN', contrary to the applicant's submission, the Board of Appeal was right to consider that it was descriptive of the goods covered by the mark applied for. As EUIPO also submitted, since that mark covers spirits in Class 33, that additional word was devoid of distinctive character and plays a very limited role in the analysis of that mark. By contrast, the most distinctive element, for the General Court, is the initial element 'PORTWO' of the same mark. Consequently, in the light of the strong similarities between the mark applied for and the PDO 'Porto', the degree of proximity of the goods at issue and the exceptional reputation of that PDO, which, moreover, is not disputed by the applicant, the Board of Appeal did not err in finding it probable that the relevant consumer would associate that mark and that PDO.[149]

Conversely. in *Portsoy*,[150] *Portus Santa Maria*[151] and *Quevedo Port*,[152] the Board of Appeal decided that there was not risk for the PDO Port as they failed to provide

[146] 14/11/2023, R 1019/2023-4, PORTE NOIRE / Port, para. 58.

[147] 21/04/2020, R 993/2019-2, Portwo gin / Porto, para. 49.

[148] 11/05/2020, R 2028/2019-2, Port ruighe / PORT/PORTO, para 55.

[149] T-417/20 - Esteves Lopes Granja v EUIPO - IVDP (PORTWO GIN), ECLI:EU:T:2021:663, para 40 to 44 and 48.

[150] Op cit 44. The PDO Port failed to provide evidence and/or at least develop a coherent line of argument to demonstrate the link between the contested sign used to designate 'Scotch Whisky complying with the specifications of the PGI Scotch Whisky' and the earlier PDO 'Port' and the Board found that the contested sign did not amount to evocation of the earlier PDO 'Port'.

[151] 14/11/2023, R 37/2023-4, PORTUS SANTA MARIA / Port, para 49. The Board of Appeal found that the relevant consumers, English-speaking or not, will not establish a link between the contested sign, 'PORTUS SANTA MARIA', applied for "tequila" (GI) agave spirit drink; "mezcal" (GI) agave spirit drink; whisky; expressly excluding wine from Jerez', and 'wine' protected by the PDO 'Porto' / 'Port' due to the significant differences in their features. Consequently, they considered that PDO Port failed to provide evidence and/or at least develop a coherent line of argument to demonstrate the link between the contested sign used to designate these non-comparable and rather distinct goods and the earlier PDO 'Porto' / 'Port'.

[152] 14/11/2023, R 2471/2022-4, Quevedo Port / Port, para. 44. The contested 'olive oils' and 'wine' protected by the earlier PDO 'Porto' / 'Port' are not comparable. The visual and phonetical similarity between the earlier PDO 'Port' and the contested sign 'Quevedo Port' is not particularly high, let alone close to identicality. From a conceptual point of view, the contested sign may even be

evidence to demonstrate the link between the contested sign and the PDO Port or on how the use of the contested sign would result in exploitation of the reputation of the earlier PDO 'Port' or because the goods were not comparable in the view of the Board of Appeal. As said by the Court in *Quevedo Port*, '*given the lack of convincing arguments or evidence to the contrary, the fact that the goods under comparison are completely different makes it very unlikely that the relevant public would make a clear and direct connection between the contested sign and the wine protected by the [earlier PDO], albeit being reputed*'.[153]

All in all, the Board of Appeal are consistent in quoting case law like *Scotch Whisky Association* and *Manchego*, where the Court has developed expansive and unassailable arguments as regards 'evocation'. Evocation can be detected even by figurative signs reproducing any graphic element of the GI's logos[154] or any type of element that could configure a connection to the GI in the consumers' mind. According to the referred Court's case-law, the concept of 'evocation' covers a situation in which the term used to designate a product incorporates part of a protected designation, so that when the consumer is confronted with the name of the product the image triggered in his mind is that of the product whose designation is protected. There is evocation of a protected name where, concerning products which are similar in appearance, the sales names are phonetically and visually similar even if the true origin of the product is indicated.[155] However, as said, it's an essential part of the Decisions of the Board of Appeal that evocation can happen although the goods are not comparable at all. In *Champanillo* (para. 66) the Court clarified that 'Evocation' does not require, as a prerequisite, that the product covered by a PDO and the product or service covered by the contested sign be identical or similar and, moreover, it is established that the use of a name creates, in the mind of an average European consumer who is reasonably well informed and reasonably observant and circumspect, a sufficiently direct and unequivocal link between that name and the PDO. The existence of such a link may arise from several elements, in particular the partial incorporation of the protected designation, the phonetic and visual similarity between the two names and the ensuing similarity, even in the absence of such elements, from the conceptual proximity of the PDO and the name in question or even similarity between the products covered by that PDO and the products or services covered by that same name.

To sum up, as describe in chapter 4 of the book, the Court of Justice has also ruled that there may be 'evocation' even if the true origin of the product is

---

perceived with a meaning which is completely unrelated to the earlier PDO 'Port'. In any event, the opponent failed to adduce evidence and/or at least develop a coherent line of argument to demonstrate specifically how the use of the contested sign, that seeks protection for 'olive oils' in Class 29, would result in exploitation of the reputation of the earlier PDO 'Port' protected for 'wine'.

[153] Op. Cit 153. Para 78.

[154] Guillem Carrau (2021), p. 391.

[155] C-44/17, SCOTCH WHISKY, EU:C:2018:415, para. 57, 60, C-87/97, Gorgonzola, EU:C:1999:115, 25, para. 46; C-4/10 & C 27/10, BNIC, EU:C:2011:484, para. 57 and 59; and C - 75/15, Viiniverla EU:C:2016:35, para. 48.

indicated[156] and even if there is no likelihood of confusion between the goods concerned,[157] since what matters is, in particular, that an association of ideas as to the origin of the product is not created in the mind of the public and that no operator takes unfair advantage of the reputation of a protected geographical indication.[158] Lastly, the Court of Justice has also stated that there may be 'evocation' of a PDO even where no protection of the European Union applies to the elements of the reference designation which reflect the terminology at issue.[159]

7.5.4 A Concept of Misleading Indications Developed in *Chenine Brut Premium Cava Bottle* (3D)

GIs are protected as well against any other false or misleading indication as to the provenance, origin, nature or essential qualities of the product that is used on the inner or outer packaging, on advertising material, in documents or information provided on online interfaces relating to the product concerned, and the packing of the product in a container liable to convey a false impression as to its origin.[160]

The Board of Appeal developed the concept of 'misleading indications' in *Chenine Brut Premium Cava Bottle (3D)*.[161] In this appeal, PDO Cava claimed that the combined use of the term 'Chenine' and the term 'Cava' in the contested trade mark was a misleading indication that was likely to mislead consumers. In their opinion, the latter might erroneously believe that the product suggested, Cava, was made using the grape variety 'Chenin blanc', which was not among the varieties established by the Cava PDO for the production of that sparkling wine, thereby creating a discrepancy between what was suggested, and the type of goods covered. However, the Board of Appeal considered that the contested sign could not be regarded as a 'misleading indication'. In particular, this conclusion was based in the following facts: no proof that a substantial part of the relevant public is familiar with the grape variety 'Chenin blanc'; no proof that the relevant public is aware of the detailed grape varieties used to produce 'Cava'; no likelihood of confusion in view of differences between the signs and low distinctive character of the word 'Cava'; and no proof of reputation of the earlier marks.

Likewise, pertinent discussion was developed in *Nero Champagne* Decision,[162] where the Board rejected the opponents' claim that the word 'Nero' was misleading in relation to wine complying with the specifications of the PDO 'Champagne' (Class 33). It noted that 'Nero' (black) would not be understood by the Italian-

[156] C-4/10 and C-27/10, Bureau national interprofessionnel du Cognac, EU:C:2011:484, para. 59.

[157] C-87/97, Gorgonzola, EU:C:1999:115, para. 26; C-132/05, Commission v Germany, EU:C:2008:117, para. 45.

[158] C-4/10 and C-27/10, Bureau national interprofessionnel du Cognac, EU:C:2011:484, para. 46).

[159] C-87/97, Gorgonzola, EU:C:1999:115, para. 26.

[160] Article 26 (1)(c) Regulation (EU) 2024/1143.

[161] Op. cit. 112, para 57.

[162] 17/02/2023, R 531/2022-2, NERO CHAMPAGNE / Champagne (GI), para. 47.

speaking consumers as an indication of the colour of the wine or of the use of black grapes only. Finally, in its *Nero Champagne* Judgment, the Court found that the word 'nero' could be perceived by consumers as evoking either the champagne grape variety or its colour, with the result that the mark applied for could convey false or misleading information.[163]

7.6 GIs Procedural Issues in EUIPO Appeal Proceedings

7.6.1 Remittal of the Case to the Examiner for Examination of the Possibility of Reopening the Examination of Absolute Grounds for Refusal: *comoengalicia.com (fig)*

It follows from Article 45(3) EUTMR and Article 30(2) EUTMDR[164] that, where the decision of the first instance of the Office is subject to appeal and the Board of Appeal considers that an absolute ground for refusal should apply to some or all of the goods or services covered by the trade mark application, the Board of Appeal may, by means of a reasoned interim decision, suspend the appeal proceedings and remit the contested application to the examiner competent for examining that application with a recommendation to reopen the examination of the absolute grounds for refusal. Where the contested application has been remitted, the examiner shall inform the Board of Appeal without delay whether the examination of the contested application has been reopened.

According to Article 30(3) EUTMDR, where the examination of the contested application has been reopened, the appeal proceedings shall be suspended until the examiner has taken a decision and, where the application is rejected in whole or in part, until the examiner's decision to that effect has become final.

In the decision *comoengalicia.com (fig)*,[165] the Board of Appeal pointed out expressly that, considering the existence of two GIs (Ternera Gallega PGI-ES-0012 and Vaca e Boi de Galicia/Vaca y Buey de Galicia PGI-ES-2308) the contested application *comoengalicia.com (fig)* in the opposition proceedings could fall under one or more absolute grounds for refusal. However, for reasons of procedural economy in this order for reference, the Board limited itself to an examination of the lack of distinctiveness under Article 7(1)(b) EUTMR.

In this context, the Board concluded in relation to the restriction of the services covered by the sign applied for, namely *retailing, wholesaling and selling via computer networks of foodstuffs from Galicia*, with the express exclusion of *Ternera Gallega y mejillón fresco de las especies Mytilus galloprovincialis cultivado en*

[163] T-239/23 | Comité interprofessionnel du vin de Champagne et INAO v EUIPO – Nero Lifestyle (NERO CHAMPAGNE) EU:T:2025:638, para. 82 to 85.

[164] Commission Delegated Regulation (EU) 2018/625 of 5 March 2018 supplementing Regulation (EU) 2017/1001 of the European Parliament and of the Council on the European Union trade mark, and repealing Delegated Regulation (EU) 2017/1430, OJ L 104, 24.4.2018, pp. 1–36.

[165] 12/02/2025, R 1291/2024-5, COMOENGALICIA.COM (fig.)/Ternera Gallega (GI) para. 13.

Sistema de bateas. (Class 35), which was accepted by the examiner, because the Board wished to state that the refusal of 'Ternera Gallega' could mean that the list of goods and services of the sign applied for was not sufficiently clear and precise, as required by the applicable case law.[166] It was therefore recommended to re-examine whether this restriction was in accordance with the law and should, where appropriate, be revoked accordingly.

The remittal for the re-examination by the Boards of Appeal can be considered an procedural practice that allow the first instance of the Office to reduce the existence of weak marks in the Registry which, at the end, will reduce the number of cancellation requests or oppositions. As regards the relationship between trade marks and GIs, the practice is relevant because, as it happened in the case at hand, will allow the first instance to reconsider the terms in which the trade mark and, where appropriate, to revoke it. This type of procedural decision, the remittal, when followed, constitutes a good example of how to find a better balance between GIs protection and trade mark owner's interest.

7.6.2 Request for an Opinion of the European Commission's DG AGRI and the Request of a GIs External Expertise: *Lucky Lasso* and *Queso y Torta de la Serena*

In *Lucky Lasso*,[167] the Board received an appeal against an Opposition division decision in which it was said that the opponent, The Distilled Spirits Council of the United States, Inc, was not allowed to take direct legal actions to prohibit the use of the trade mark 'Lucky Lasso Kentucky Straight Bourbon Whiskey' because the applicable legal framework had not specific rules for the protection to permit it in an international agreement invoked by the opponent.

The international agreement was an exchange of letters between the USA and EU in 1994 and contained the appellation 'Bourbon'. The Board of Appeal sought and received advice from the European Commission, Directorate-General for Agriculture and Rural Development (DG AGRI). DG AGRI confirmed that the agreement, in the form of an exchange of letters of 25/03/1994, had not been denounced by either party and therefore remained in force and that:

> In any event the agreement in the form of an exchange of letters is not selfexecuting and its current implementation is through Commission Regulation (EC) No 936/2009 of 7 October 2009 applying the agreements between the European Union and third countries on the mutual recognition of certain spirit drinks.
>
> …
>
> The annex confirms that all the terms "Bourbon Whisky", "Bourbon Whiskey" and "Bourbon" where it is used to describe Bourbon Whiskey are protected names. The level of protection is that applied in the EU's domestic Spirits GI regime under Regulation (EC) No

[166] T-432/16, меprivilege privilege prehensusta, EU:T:2017:527, para. 46.

[167] 23/04/2019, R 2723/2017-5, Lucky Lasso Kentucky Straight Bourbon Whiskey / Protected designation of origin or a geographical indication Bourbon Whiskey.

110/2008 of the European Parliament and of the Council of 15 January 2008 on the definition, description, presentation, labelling and the protection of geographical indications of spirit drinks and repealing Council Regulation (EEC) No 1576/89.

...

The scope of protection is set out in Article 16 and the relation with trade marks in Article 23. Article 24 refers to controls exercised by Member States. We see no ground to limit the scope of protection applied to these terms, which is that laid down in Regulation (EC) No 110/2008. The reference to the laws of 3rd countries in Commission Regulation (EC) No 936/2009 applies to the composition and origin of the product, while the protection and labelling rights and prohibitions within the EU are laid down in EU law.[168]

Following DG AGRI's advice, the Board considered that the agreement was not self-executing and required an expression in EU Law that was provided by was provided by Regulation (EC) 1267/94 citing Regulation (EEC) 1576/89. Then, it happened that Regulation (EC) 1267/94 was repealed and replaced by Regulation (EC) 936/2009, which cites the successor to Regulation (EEC) 1576/89, that is, Regulation (EC) 110/2008 of the European Parliament and the Council. The last one contained an article that regulates the use of the GIs and states, inter alia, that the registration of a trade mark shall be refused or invalidated where it contains or consists of a GI if its use would lead to any of the situations referred to in Article 16 Regulation (EC) 110/2008.[169] From June 2019, the provisions of Regulation (EC) No 110/2008 have been replaced by Regulation (EU) 2019/787.[170]

The Board concluded that there was a direct reference to the relevant goods and clearly fell under the referred Article 16 in that the sign evoked Kentucky whiskey, making a direct reference to bourbon sourced from the USA in general and that state in particular. Therefore, the Board considered that the mark must be refused for registration based on Article 25 Regulation (EC) 110/2008 and Article 8(6) EUTMR an annulled the contested decision.

In *Queso y Torta de la Serena*,[171] the Court annulled the Board of Appeal decision on the case at hand[172] because infringed, on the one hand, Article 2(2) and Article 3(1) of Regulation No 510/2006, by not examining whether the term 'Torta del casar' was a traditional non-geographical name, and, on the other hand, Article 13(1), first subparagraph, (b), of that regulation, by ruling out the possibility that the

---

168 Para. 8 of the referred Decision.

169 Regulation (EC) No 110/2008 of the European Parliament and of the Council of 15 January 2008 on the definition, description, presentation, labelling and the protection of geographical indications of spirit drinks and repealing Council Regulation (EEC) No 1576/89, OJ L 39, 13/02/2008, pp. 16–54.

170 Regulation (EU) 2019/787 of the European Parliament and of the Council of 17 April 2019 on the definition, description, presentation and labelling of spirit drinks, the use of the names of spirit drinks in the presentation and labelling of other foodstuffs, the protection of geographical indications for spirit drinks, the use of ethyl alcohol and distillates of agricultural origin in alcoholic beverages, and repealing Regulation (EC) No 110/2008, OJ L 130, 17.5.2019, pp. 1–54.

171 T-828/16, Consejo Regulador "Torta del Casar" v EUIPO - Consejo Regulador "Queso de La Serena" (QUESO Y TORTA DE LA SERENA) ECLI:EU:T:2017:918, para. 64 and 66.

172 26/09/2016, R 2573/2014-4, QUESO Y TORTA DE LA SERENA (FIG.) / TORTA DEL CASAR et al.

trade mark applied for evoked the PDO 'Torta del casar' simply because the term 'Torta' did not indicate a geographical location.

On 23 April 2018, the Registrar of the Boards informed both parties that the appeal had been reassigned to the Fifth Board of Appeal. On 4 September 2018, the Fifth Board contacted the European Commission, Directorate-General for Agriculture and Development, seeking an opinion on whether or not the term 'Torta' contained in the protected designation of origin (PDO) 'Torta del Casar' was generic and whether the trade mark 'Queso y torta de la Serena' evoked the PDO 'Torta del Casar'.[173]

DG AGRI replied first that, as regards the term 'Torta' and its 'genericity' that Article 3 of Regulation (EU) No 1151/2012 stipulates that 'generic terms' are 'names of products which, although referring to the place, region or country where a product was originally produced or marketed, have become the common name of that product in the Union'. The term 'Torta' comes from the unusual form of cheese, due to its specific method of curing. In our opinion, it does not have a geographical connotation per se and cannot therefore be described as generic in the terms of Regulation (EU) No 1151/2012. However, DG AGRI said that it could be considered as a common noun (like 'ham', 'chicken', 'meat' or 'provola'). The Commission has no means to judge whether the term 'Torta' is a common name when used in the name of a Spanish cheese. In order to give this opinion it would be necessary to analyse the historical, cultural and linguistic context in which the word is used. If it could be concluded that the term 'Torta' used for a Spanish cheese is a common noun, as is the case with the system applicable to generic names (see Article 13(1) last paragraph of Regulation (EU) No 1151/2012), the protection granted by Article 13 to the name 'Torta del Casar' could not be extended to that part of the name corresponding to a common name (torta) and therefore said term would be freely used.

Second, concerning the evocation and the legitimation to market cheese in the form of a pot, DG AGRI referred to the single document to identify the shape as a specific characteristic in the case of 'Torta del Casar' in the part describing the specificity of the product and the part containing the description of the causal link. Likewise, DG AGRI said that '*a comparison between the two PDOs leads one to ask about the potential likelihood of confusion on the part of the consumer and the likelihood of evocation, given the geographical proximity of the respective production areas of the two PDOs (located in Extremadura), the identity of the class (cheeses) and the similarity of the nature of the goods (sheep's cheese: 'Serena cheese' is a semi-hard, ripened soft cheese made from sheep's milk from the Merina/'Torta del Casar' breed is a cheese made from raw sheep's milk from the logs 'Merino' and 'entrefino').*'

The opponent had the opportunity to submit observations on the questions put by the Board of Appeal to the European Commission, as well as on the answers given

[173] See para 16 of 25/02/2021, R 696/2018-5, QUESO Y TORTA DE LA SERENA (fig.) / Torta del Casar et al.

by that body. DG AGRI answers are quoted several times in the reasoning of the Decision.[174]

The request for external expertise on GIs was a *never-before-seen* in appeal proceedings in the field of GIs. The decision contains a brief summary of all the procedural steps adopted by the Board of Appeal to get this external expertise on GIs: firstly, communication to the parties indicated that the Board of Appeal started seeking expert evaluations and appoint experts to determine the cultural, historical, economic and social indexes and suspension until all the expert opinions have been brought to the knowledge of the parties for their observations; secondly, communication to the parties was informing them that the Board, in accordance with the above, had decided to appoint the expert to answer a number of questions; thirdly, a provisional decision with the request for expertise; fourthly, the report was received and it contained the answer to the question and the proposal of a demoscopic study. Finally, the applicant submitted observations on the opinion of the expert, then decided to withdraw them and therefore did not appear in this decision. The opponent sent comments after the expert's report.

As an outcome of the request, it must be outlined that the expert's report is quoted several times in the reasoning of the Decision[175]; in particular, in the paragraphs on the possible generic nature of the term 'Torta' (para. 51) when it is said that '*...This Board is of the opinion that the term 'Torta' is not generic. This conclusion is apparent from a series of indexes studied in the expert's report, which is explained*' and on the use in the course of trade of more than mere local significance with the conclusions and the data produced by the expert on the basis of the information published by the Spanish Ministry of Agriculture (para. 99).

7.6.3 Substantiation of the GI in Opposition Proceedings: From *Grilloumaki* to *CampoVerde*

Article 7(2)(e) EUTMDR is applicable to oppositions based on Article 8(6) EUTMR and stipulates that:

> (e) where the opposition is based on an earlier designation of origin or geographical indication within the meaning of Article 8(6) of Regulation (EU) 2017/1001, evidence of its acquisition, continued existence and scope of protection including, where the earlier designation of origin or geographical indication is invoked pursuant to the law of a Member State, a clear identification of the content of the national law relied upon by adducing publications of the relevant provisions or jurisprudence

Therefore, the opponent must file evidence of the existence, validity and scope of protection of its earlier GI, as well as evidence of its entitlement to file the opposition.

---

[174] 25/02/2021, R 696/2018-5, QUESO Y TORTA DE LA SERENA (fig.) / Torta del Casar et al. Para 49 to 50.

[175] 25/02/2021, R 696/2018-5, QUESO Y TORTA DE LA SERENA (fig.) / Torta del Casar et al. Para 51 and 99.

Arguably, the substantiation of the earlier GI was not an easy sailing question according to the requirements described by EUIPO Guidelines. EUIPO Guidelines explained that pursuant to Article 7(2) EUTMDR, the opponent must submit evidence of the existence and scope of protection of the earlier GI, as well as evidence of its entitlement to file the opposition. This applies to all actions, regardless of whether they are based on EU GIs, GIs protected by the laws of Member States or GIs protected by international agreements. According to Article 7(2)(e) EUTMDR, in order to substantiate its right, the opponent must provide the Office with evidence of the existence and scope of protection of the GI invoked. The opponent must also prove its entitlement to file the opposition, namely, that it is authorised under the relevant law to exercise the rights arising from a GI (Article 46(1)(d) EUTMR and Article 7(2) EUTMDR). Furthermore, the opponent had to prove that the applicable law confers on it a direct right of action to prohibit unlawful use of a GI. From July 2026, a new version of the EUIPO Guidelines facilitates the substantation of EU GIs in Article 8(6) EUTMR because EUIPO has *ex officio* knowledge of the Union Registers (Union register of GIs for wines, spirit drinks and agricultural products, Union register of GIs for craft and industrial products) and the substantation may be based on the entry in the Union Register of the recognised producer group of a GI (GIs for wine, spirits and agricultura products).[176]

Therefore, until July 2026, to substantiate its right, the PDO/PGI had to provide the Office with the necessary facts and evidence regarding the validity and the scope of protection of its right. Depending on the GI, documents such as publication and registration in the Official Journal or an extract from the official register were deemed sufficient if they contained sufficient data to determine all the relevant particulars of the earlier right (e.g., protected name, application or registration date, goods protected by the GI).

Given that, pursuant to Article 95(1) EUTMR, in proceedings relating to relative grounds for refusal of registration, the Office is restricted in its examination to the facts, evidence and arguments submitted by the parties and the relief sought, the existence of these conditions must be proven by the opponent within a time limit specified by the Office, according to Article 7(1) EUTMDR.

On the one hand, in *Grilloumaki*,[177] the Board annulled the contested decision as the producers of the PDO are allowed to file an opposition based on Article 8(6) in conformity with Regulation (EU) 2024/1143. The Board of Appeal decided on the entitlement to file the opposition by the 'Group of processors and producers of sheep milk and goatmilk' by concluding that the applicable law, in particular Article 55 read in conjunction with Article 2 of the Regulation (EU) 2024/1143 (then Article 45 and 3 of Regulation (EU) No 1151/2012), enables any group of producers or processors working with the corresponding product to take actions to ensure adequate legal protection of the PDO as for example filing an opposition.

---

[176] See EUIPO Guidelines, Part C, Opposition, Section 6, Geographical Indications, paragraphs 2.4 Specifics of substantiation and 3.1.3. Limits to the scope of protection of GIs on relative grounds.

[177] 29/08/2022, R 462/2022-5, GRILLOUMAKI / Χαλλουμι' (Halloumi)/'Hellim. para. 16.

The Board considered that it must apply even more so if the group is the 'applicant group' in the sense of Article 56 of the Regulation (EU) 2024/1143. The applicable law defines the concept of 'group of producers and processors' very widely by referring processors and producers of sheepmilk and goatmilk'. In the case at hand, it was not a legal entity as such but merely a group of nine companies which produce or process sheep milk and goatmilk, they are a group in the sense of Article 3(2) of Regulation (EU) 1151/2012. As 'a group', they comply with the conditions laid down in Article 2 Regulation (EU) 2024/1143 and are entitled to file the present opposition according to Article 55 Regulation (EU) 2024/1143. In this sense, as the opponent correctly emphasises, the rules on the entitlement to take actions to protect a PDO as contained in Regulation (EU) 2024/1143 are lex specialis to Article 3 EUTMR. The case was remitted to the Opposition Division to examine the existence of evocation considering that that the Board's findings on the entitlement of the opponent are binding upon the Opposition Division.

In *Macroom Buffaloumi*,[178] the Board concluded that the contested decision erred in finding that the 'Group of processors and producers of sheep milk and goat milk' was not entitled to file the opposition. According to the Article 55 and 2 of the Regulation (EU) 2024/1143 (in the applicable law at that time Article 45 and 3 of Regulation (EU) No 1151/2012), 'any group' of producers o processors, irrespective of its legal form, is entitled to defend a PDO, to the extent that they produce or process the corresponding foodstuffs or agricultural goods.

Moreover, the Board have been consistent on the subject matter of the substantiation of the earlier right in Halime (fig)[179]; Romitòro (fig) and Romitoro (fig),[180] and Bolgaré (fig)[181] where the cases were remitted to the Opposition Division for a full assessment of the opposition; or to renew the assessment of substantiation like in Teslaquila.[182]

On the other hand, in *Campo Verde*,[183] the Board of Appeal concluded that the opposition based on the right PDO Vinho Verde was rightly rejected as unfounded. The opponent ought to have produced evidence that it was entitled to file the opposition based on such earlier right. However, such evidence was not submitted. The Board noted that the Opposition Division rightly pointed out that, according to Article 95(1) EUTMR (which in the part to be referred to is identically worded to Article 76(1) EUTMR in the version in force at the time of commencement of the adversarial part of these opposition proceedings), in proceedings relating to relative

[178] 29/07/2022, R 1940/2021-5, Macroom buffaloumi / Halloumi, para. 24.

[179] 25/07/2022, R 1939/2021-5, Halime (fig.) / 'ΧΑΛΛΟΥΜΙ' (HALLOUMI)/'HELLIM'.

[180] 11/04/2022, R 100/2022-5, ROMITÒRO (fig.) / Toro and 04/04/2022, R 101/2022-5, ROMITORO / Toro. The Board noted that, taking into account the document filed at first instance, together with the additional documents filed at the appeal stage, the opponent has properly substantiated its earlier right, pursuant to Article 7(2) EUTMDR. The case is remitted to the Opposition Division for a full assessment of the opposition.

[181] 21/03/2022, R 2564/2019-2, BOLGARÉ (fig.) / Bolgheri et al.

[182] 08/03/2022, R 803/2021-2, Teslaquila / TEQUILA (GI).

[183] Op cit para. 93 to 111.

grounds of refusal of registration, the Office shall be restricted in this examination to the facts, evidence and arguments provided by the parties and the relief sought. Considering the applicable rules, only an extract of the relevant information from the EU register of PDO and GI for wine may show evidence of the existence and validity of the PDO 'Vinho Verde'.

Similarly happened in *Delicius Green Eva Fricke*[184] as PDO Vinho Verde relied on the national registrations of the GI for wines, which since the uniform protection exist in the EU cannot invoked as basis of the opposition under Article 8(6) EUTMR. Although new evidence was presented before the Board, in particular an extract of from the European Commission Register, the PDO 'Vinho Verde' did not submit any valid information on the legal protection of the three PDOs invoked in its notice of opposition. The proof of the relevant registration of the PDO was submitted for the first time before the Board. Since the opponent did not submit, in the first-instance proceedings, any evidence of the EU-level registration of its PDO, the documents filed for the first time at the appeal stage are entirely new and not merely supplementary to evidence adduced in due time.[185]

In the context of substantiation of rights of GIs protected under international agreements between the EU and third countries or under bilateral international agreements with third countries signed by a Member State, the Boards of Appeal have adopted a consistent approach that can be explained with the following examples:

- *Lucky Lasso*.[186] The Board decided on a case relating to a EUTM consisting, amongst others, of two elements of a GI for bourbon and whiskeys. Such GI enjoyed protection because of a bilateral agreement between the United States and the EU, made in the form of an exchange of letters. The Board, as referred before, during the proceedings asked for the opinion of the European Commission (DG Agri) about the effective protection of the GI and the validity of the agreement. Even if the agreement was not self executing, it was considered to be implemented in the EU system through the 'Commission Regulation (EC) No 936/2009 of 7 October 2009 applying the agreements between the European Union and third countries on the mutual recognition of certain spirit drinks'.
- *Matusalem el Espíritu de Cuba*.[187] In this case, the opposition was based on Article 8(4) EUTMR and on the use made of the Geographical Indication 'Cuba' for 'rum' protected under bilateral agreements signed between the Kingdom of Spain and the Republic of Cuba on 23 January 1979 and between the Federal Republic of Germany and the Republic of Cuba on 22 March 1954. Despite the numerous documents filed, such as the international agreement between Spain and Cuba, Spanish judgements, Spanish administrative decisions, Cuban legisla-

[184] Op cit para. 12 to 32.

[185] 15/09/2020, R 2931/2019-1, Eldstickan / SOLSTICKAN (fig.) et al., § 41.

[186] 23/04/2019, R 2723/2017-5, Lucky Lasso Kentucky Straight Bourbon Whiskey / Protected designation of origin or a geographical indication Bourbon Whiskey.

[187] 05/02/2018, R 1108/2017-2, Matusalem Espíritu de Cuba/Cuva.

tion, etc., the opponent failed to prove the rule in which the internal Spanish legislation confers the right to prohibit use of latter trade marks and the conditions under which this right can be exercised. As to the right to prohibit use in Germany, the provision pointed out by the opponent in the agreement between the Federal Republic of Germany and the Republic of Cuba on the reinstatement of Industrial Property Rights and Geographical Indications of 22 March 1954 and Section 127 of the German Act on the Protection of Trademarks and Other signs do not provide for a subjective right granted to a party who, being the owner of a Cuban GI, under specified conditions can prohibit use of a subsequent trade mark application.

- *Café de Cauca.*[188] In the case at hand, the opposition was based on the GI 'Café de Cauca' for coffee protected in the European Union under the Trade Agreement between the European Union and its Member States, of the one part, and Colombia and Peru, of the other part.[189] By decision of the Commission No 2018/1733 of 14 November 2018 the Agreement was approved on behalf of the European Union. 'Café de Cauca' is protected as a GI in the EU as published in the Official Journal on 16/11/2018. It is included in GI View.

7.6.4 Right to Be Heard: *PriSecco*

In *PriSecco*,[190] we can find an example of a procedural violation as regards the right to be heard principle in GIs proceedings. In the case at hand, the producers of the PDO *Prosecco* filed a request for a declaration of invalidity of 'PriSecco' based on Article 59(1)(a) EUTMR in conjunction with Article 7(1)(g) EUTMR, and Article 60(1)(d) EUTMR in conjunction with Article 8(6) EUTMR. The EUTM application was for Class 32: cocktails, non-alcoholic.

The Cancellation Division declared the invalidity of the contested EUTM for all the contested goods and the decision was appealed before the Board. A little bit later, the EUTM proprietor submitted a declaration of partial surrender requesting that the list of goods be restricted as follows: Class 32: non-alcoholic cocktails whose ingredients are apple and or pear juices based on meadow fruit varieties (WiesenObst). As it can be gathered from paragraphs 11 and onwards of the quoted Decision, the Office informed the EUTM proprietor that there were deficiencies in the declaration of partial surrender and proposed it that the list of goods be amended as follows: Class 32: non-alcoholic cocktails whose ingredients are apple and or pear juices based on meadow fruit varieties. Following the acceptance of the proposal by the proprietor, the Registry of the Board of Appeal directly confirmed the acceptance of the partial surrender and invited the producers of the PDO Prosecco

[188] 13/10/2023, R 825/2023-5, CAVCA / Café de Cauca (GI).

[189] OJ L 354, 21.12.2012, p. 3.

[190] 20/02/2024, R 1454/2022-5, PriSecco / 'Prosecco' para. 93–94.

to confirm its interest to proceed in the invalidity proceedings. The PDO Prosecco confirmed its legal interest to obtain a declaratory decision.

The Board of Appeal's rapporteur considered that the decision of the Registry was taken without affording the producers of the PDO *Prosecco* an opportunity to be heard first, constituted a clear procedural violation. It infringed upon the PDO Prosecco's right to be heard before accepting the partial surrender requested by the EUTM proprietor.

The invalidity applicant based its claim concerning the existence of a legitimate interest in obtaining a decision on the merits (also) in relation to the surrendered goods essentially on two arguments:

1) *The preliminary injunction (PI) application filed on 23 September 2020, before the Court of Milan against, inter alia, the EUTM proprietor.*
2) *The fact that, in the present cancellation proceedings, the invalidity applicant also invoked grounds related to the 'public interest', namely that the challenged EUTM was registered in breach of Article 7(1)(g) EUTMR and Article 103(2) of Regulation (EU) No 1308/2013, letters (c) and (d).*

Both parties acknowledged that the proceedings before the Court of Milan arising from the preliminary injunction have been terminated and that there were currently no other parallel proceedings between the parties concerning the contested EUTM.

It is devoid of any basis since its acceptance would lead to the consequence that, in all requests for invalidity based on absolute grounds (as there are public interest reasons behind all of them), the invalidity applicant would be relieved of the duty to justify the existence of a legitimate interest in obtaining a decision on the merits, despite the total or partial surrender of the contested EUTM.

The Board of Appeal concurred with the rapporteur that the registration of the partial surrender should have been suspended, and the producers of the PDO Prosecco as invalidity applicant should have been invited to file observations thereon. On the top of that, the producers of the PDO Prosecco had a legitimate interest in obtaining a decision on the merits. Consequently, the Board hereby decided to revoke the acceptance of the partial surrender of the contested EUTM on the basis of a breaching of the right to be heard of the PDO Prosecco.

All in all, the effects of partial surrender during the cancellation proceedings are ex nunc according to Art. 57(2) EUTMR. The Board of Appeal stated that it is for the invalidity applicant to demonstrate a real, direct, and present legitimate interest in obtaining a decision on the merits also in relation to the surrendered goods, so that to have effects of ex tunc under Article 62(2) EUTMR. Moreover, this litigation highlights how pertinent is to involve both parties at any step in the proceedings before the Board of Appeal.

Therefore, the application for invalidity was upheld in its entirety due to a conflict with the PDO *Prosecco* because, at least from the perspective of a non-negligible part of the relevant EU public, the contested EUTM did constitute an evocation of the earlier PDO.

7.6.5 Submission of Belated Evidence: Vega del Obispo Jaén Productos Ecológicos

In *Vega del Obispo Bio Jaén Productos Ecológicos* (fig),[191] the Decision of the Board of Appeal was annulled by the General Court[192] because it has been taken in the absence of a reason justifying the late submission of evidence. Therefore, the second condition of Article 27(4) EUTMDR was considered not fulfilled.

The Court recalled that, according to Article 95(2) EUTMR, EUIPO 'may disregard facts which the parties have not raised or evidence which they have not produced in good time'.[193] For this reason, it was considered that by specifying that EUIPO 'may', in such a case, decide not to take such evidence into account, Article 95(2) of Regulation 2017/1001 in fact grants it a broad discretion to decide, while giving reasons for its decision on this point, whether to take such evidence into account.[194]

The possibility for the Board of Appeal to consider evidence submitted for the first time before it, pursuant to Article 95(2) of Regulation 2017/1001, is governed by two cumulative conditions provided for in Article 27(4) EUTMDR. The General Court, in paragraphs 22 and 23 of the judgment stipulates that the evidence might be accepted if it could be relevant to the outcome of the case and if it was not submitted in good time for valid reasons; when the belated evidence adds relevant facts and evidence that had already been submitted in good time, or is filed to challenge the conclusions drawn or examined ex officio by the first instance in the decision under appeal.

> 22 Indeed, according to that provision, the Board of Appeal may accept evidence produced for the first time before it only if that evidence meets two conditions. First, if it appears, prima facie, to be relevant to the outcome of the case. Second, if it was not submitted in good time for valid reasons, in particular where it merely supplements relevant facts and evidence that had already been submitted in good time, or is filed to challenge the conclusions drawn or examined ex officio by the first instance in the decision under appeal.
>
> 23 Furthermore, it should be noted that it is for the party submitting evidence for the first time before the Board of Appeal to explain to it the extent to which that submission satisfies the conditions laid down in Article 27(4) of Delegated Regulation 2018/625 (see judgment

[191] 01/06/2023, R 1119/2022-1, VEGA DEL obispo BIO Jaén organic PRODUCTOS (fig.) the Board considered that the contested mark was not excluded from registration, pursuant to Union legislation or national law or to international agreements to which the Union or the Member State concerned is party, providing for protection of PDO and PGIs according to Article 7(1)(j) EUTMR.

[192] T-425/23, Consejo Regulador «Aceite de Jaén» v EUIPO - Agrícola La Loma (VEGA DEL OBISPO BIO Jaén PRODUCTOS ECOLÓGICOS), para. 18–28.

[193] It follows, states the Court, from the wording of that provision that, as a general rule and unless otherwise provided, the submission of facts and evidence by the parties remains possible after the expiry of the time limits to which such submission is subject, pursuant to the provisions of that regulation, and that EUIPO is in no way prohibited from taking into account facts and evidence thus submitted or produced late (see judgment of 2 June 2021, Franz Schröder v EUIPO – RDS Design (MONTANA), T-854/19, EU:T:2021:309, paragraph 24 and the case-law cited).

[194] T-854/19, Montana EU:T:2021:309, paragraph 25 and the case-law cited.

of 13 September 2023, Prolactal v EUIPO – Prolàctea (PROLACTAL), T-549/22, not published, EU:T:2023:538, paragraph 28 and the case-law cited).

In this context, the Court considered appropriate to examine whether, in exercising its discretion, the Board of Appeal complied with the cumulative conditions established by Article 27(4) EUTMDR and noted that, in the contested decision, the Board of Appeal made no reference to the reasons why it admitted the evidence produced by the proprietor of the contested mark for the first time before it.[195]

7.7 Collective Trade Marks and GIs: The *Darjeeling tea* Case and the Weakening of GIs

With roots in the mandatory guild mark, collective marks are now voluntary and serve not only a distinctive purpose, but also a significant promotional or competitive function.[196] Among the key areas for potential growth of these signs, the increasing commercial appeal of certain geographic origins is notable, often linked to the production of specific products. This has led to the emergence of geographic collective marks, which combine geographic indications with generic terms to create a distinctive identity.[197]

Collective marks are owned by a collective body such as, for example, a trade association or an association of producers or manufacturers and serve to indicate that the person who uses the collective mark is a member of that collectively. Membership in the association that is the owner of the collective mark is subject to compliance with certain rules, such as the geographical area of production of the goods on which the collective mark is used, or standards of production of such goods.[198]

There is a need of principles and rules to solve the conflicts that could arise in situations where a collective trade mark linked to the production of specific products must coexist with a previous GI in the market.[199]

As the Court of Justice declared in *Darjeeling*[200] the essential function of an EU collective trade mark consists of distinguishing the goods or services of the

[195] See para. 25 to 27 of the quoted Judgment where the Court says that '25. In the present case, on 27 September 2022, EUIPO notified the applicant's statement of grounds on the proprietor of the contested mark, indicating to it that the deadline for filing observations in response was 27 November 2022. It was only on 24 April 2023 that the proprietor of the contested mark sent the certification to the Board of Appeal, without giving any reason justifying its submission at a late stage in the proceedings before the Board of Appeal.

[196] Baylos Corroza (1993), p. 843.

[197] Largo Gil (2006), p. 314 and De Martín Muñoz (2003), p. 143.

[198] Op. cit. WIPO page 11.

[199] Medina González (2012), p. 251.

[200] C-675/15 P (Joined Cases C-673/15 P, C-674/15 P, C-675/15 P, C-676/15 P) The Tea Board v EUIPO, ECLI:EU:C:2017:702, para. 43.

members of the association which is the proprietor of the mark from those of other undertakings and not according to their geographical origin:

> 43 Consequently, since none of the provisions in the chapter of Regulation No 207/2009 dedicated to Community collective marks allows it to be inferred that the essential function of such marks, including those consisting of an indication which may serve to designate the geographical origin of the goods covered, is different from the essential function of Community individual marks, it must be held that the essential function of the former — as it is for the latter — is to distinguish the goods or services covered according to the specific body from which they originate and not according to their geographical origin.

The EU legislator considers that it is possible to register geographical indications as EU collective trade marks. As from Article 74 EUTMR, by way of derogation from Article 7(1)(c), signs or indications which may serve, in trade, to designate the geographical origin of the goods or services may constitute EU collective marks. Consequently, there is room to register GIs as collective trade marks. However, as stated by the Court in *C-766/18 Halloumi,* collective trade marks are not exempted to the requirement of distinctiveness: '*73 Article 66(2) of Regulation No 207/2009 is not an exception to that requirement of distinctiveness. While that provision permits, by way of derogation from Article 7(1)(c) of that regulation, registration as EU collective marks of signs which may serve to designate the geographical origin of goods or services, it does not, on the other hand, allow the signs thus registered to be devoid of distinctiveness. Where an association applies for registration, as an EU collective mark, of a sign which may designate a geographical origin, it is therefore incumbent on it to ensure that sign has elements which enable the consumer to distinguish the goods or services of its members from those of other undertakings*'.

In any case, there is a general perception on the fact that collective trade marks containing a GI are weak marks and, therefore, their defence in opposition proceedings is complicated. This perception has been enriched from the saga of cases on *Halloumi.*[201] Likewise, the Decisions of the Board of Appeal contribute, except for Riojavina,[202] to it. If anyone whose products originate from the geographical area can become a member of the association owning the EU collective trade mark which reflects or contains a GI, the GIs *ius prohibendi* is limited.[203]

On the top of that, EU collective trade marks are under a regime of genuine use and renewal as an EUTM, which is not the case for GIs. In the case known as *Sterisches Kürbiskernöl,*[204] a cancellation action opened a debate on the genuine use of the figurative mark for *Pumpkin seed oil produced according to the specifications of the mentioned GI.* The findings of the Court were that it must be considered that the use of the mark does not allow the goods of an undertaking which are covered by that mark to be distinguished from the goods which are covered by that mark but

---

[201] Among other, for instance, T-534/10, HELLIM / HALLOUMI, EU:T:2012:292. And C-393/12P, HELLIM / HALLOUMI, EU:C:2013:207.

[202] T-138/09, RIOJAVINA, EU:T:2010:226.

[203] Largo Gil (2006), p. 330.

[204] T-72/17, Gabriele Schmid v. EUIPO, ECLI:EU:T:2018:335, para. 56.

which come from other undertakings. In other words, the use of a GI logo does not serve as use of a trade mark.

In any case, following the Court in Halloumi case, collective marks consisting on a GI are excluded from. In that regard, the General Court has already held that registration of a sign as a collective mark cannot, per se, give rise to a presumption that the mark has average distinctive character (judgment of 13 June 2012, Organismos Kypriakis Galaktokomikis Viomichanias v OHIM – Garmo (HELLIM), T-534/10, EU:T:2012:292, paragraph 52). In this respect, it must also be held, as regards the need to protect EU collective marks which, under the provisions of Article 66(2) of Regulation No 207/2009, are expressly entitled to designate the geographical origin of the goods or services to which they relate, that the Court of Justice noted, in paragraph 74 of the judgment of 5 March 2020, Foundation for the Protection of the Traditional Cheese of Cyprus named Halloumi v EUIPO (C-766/18 P, EU:C:2020:170), that, even supposing that the EU collective mark HALLOUMI implicitly refers to the Cypriot geographical origin of the goods covered, it must nevertheless still fulfil its essential function, namely to distinguish the goods or services of the members of the association which is the proprietor of that mark from those of other undertakings. In the present case, the generic nature of the word 'halloumi', since that word alone constitutes the earlier mark, necessarily limits the effects of the mark as regards that function, whatever the descriptive character of that mark from a geographical point of view.[205]

7.8 GIs Versus EU Certification Marks

Certification marks are defined as marks that are registered and used for the specific purpose of guaranteeing that the products or services labelled or designated with them meet certain standards, possess specific qualities or characteristics, or have a particular origin.[206]

Article 15(2) Directive 2008/95/EC[207] recognise a margin of manoeuvre to Member States by allowing them to create, at national level, guarantee or certification trade marks, that indicate geographical origin. It stipulates that:

> By way of derogation from Article 3(1)(c), Member States may provide that signs or indications which may serve, in trade, to designate the geographical origin of the goods or services may constitute collective, guarantee or certification marks. Such a mark does not entitle the proprietor to prohibit a third party from using in the course of trade such signs or indications, provided he uses them in accordance with honest practices in industrial or commercial matters; in particular, such a mark may not be invoked against a third party who is entitled to use a geographical name.

[205] T-106/22, Bbqloumi / Halloumi, para. 55.

[206] Baylos Corroza (1993), p. 847 and Montero García-Noblejas (2023), p. 89.

[207] Directive 2008/95/EC of the European Parliament and of the Council of 22 October 2008 to approximate the laws of the Member States relating to trade marks, *OJ L 299, 08/11/2008.*

Conversely, the same EU legislator, according to Article 83(1) EUTMR, has deliberately excluded the geographical origin element from the scope of EU certification trade marks, as follows:

> An EU certification mark shall be an EU trade mark which is described as such when the mark is applied for and is capable of distinguishing goods or services which are certified by the proprietor of the mark in respect of material, mode of manufacture of goods or performance of services, quality, accuracy or other characteristics, with the exception of geographical origin, from goods and services which are not so certified.

Regarding certification marks, it should be noted that the nature of the signs that can became registered as a GI is different from that of the signs that can be configured as a trademark. While the signs that can form a GI can only be composed of a geographic name, the signs that constitute trademarks can be very varied. Additionally, although both institutions are indicative of origin, the GI is only indicative of the geographic origin of the products, whereas the trademark is indicative of their business origin.[208] On the other hand, collective and guarantee marks have a difference between the owner and the user of the mark; collective and guarantee marks have a determined period of time for their use; and, finally, collective or guarantee marks can use words or combinations of words, images, figures, symbols, graphics, or any combination of signs or means to distinguish their products in the market without needing to be geographic.[209] All of this without forgetting that the essential difference lies in the fact that designations of origin are marked by public law from their inception.[210]

Ultimately, a fundamental distinction exists between these two types of intellectual property rights, certification marks provide direct protection to the consumer, whereas GIs are rooted in the interests of the producer.[211]

7.9 Conclusions

The EU GIs scheme shows the complex relationship between GIs and trade marks. It has reached a level of sophistication where 'coexistence' is often synonymous with 'calculated tension'. Based on the detailed analysis of the decisions of the EUIPO Board of Appeal in absolute ground and in inter partes proceedings, this chapter has developed conclusions on the scope and extension of GI protection in trade mark disputes, highlighting both the strengths and vulnerabilities of GIs, including their interplay with collective and certification trade marks. It is not merely a technical conflict of laws; it is a profound legal dialogue between collective territorial heritage and individual commercial identity. The analysis of the Board of Appeal decisions seeks to provide an appropriate reference of decisions,

[208] Botana Agra (2001), p. 26.

[209] Leach Ros (2002), p. 325.

[210] López Benítez (1996), p. 75.

[211] De la Calle Robles (2002), p. 31.

including *Campo Verde, Toro de Piedra*, and *Chenine Brut Premium Cava Bottle, Colombiano Coffee House, Romeral de Jabugo, Padrón Auténtico and Marqués de la Mancha*. Through these key decisions, the Boards of Appeal answer the questions related to GIs protection in trade mark disputes, providing insights into the interactions between these two intellectual property rights. The analysis of Article 7(1)(j) EUTMR reveals that the EUIPO has moved toward a proactive stance in preventing the 'dilution by registration' of GIs. The application of the *BNIC Cognac* criteria and the evolving concept of evocation (as seen in the *Padrón Auténtico* line of cases) suggest that the "comparability" of goods is being interpreted broadly to safeguard the reputation of the GI. This prevents applicants from finding "safe positions" through clever limitations of their goods and services, ensuring that a GI's prestige cannot be easily circumvented by proximity. A central finding of this research is the paradox of GI protection in inter partes proceedings. While cases like *Campo Verde* and *Toro de Piedra* highlight the BoA's commitment to preventing direct and indirect misuse, they also expose a significant vulnerability: the difficulty of demonstrating a likelihood of confusion. Here comes the 'evocation' shield: from *Romeral de Jabugo* to *Marqués de La Mancha* underscores that 'evocation' remains the most potent weapon for GI holders, often providing a wider scope of protection than traditional 'likelihood of confusion' metrics. Besides, the procedural analysis highlights that the 'effective advocacy' of a GI could be often won or lost on technicalities. The chapter concludes that a GI's legal strength is only as effective as its procedural defense; failures in providing evidence of the GI's scope or belated submissions can affect the GI protection. Finally, the interplay between GIs and other collective IP rights presents a cautionary tale or a landmark warning: the "weakening" of a GI occurs when its unique territorial character is subsumed by the functional characteristics of a Collective Trade Mark. Furthermore, the emergence of EU Certification Marks introduces a new competitor in the "quality signaling" market. This chapter concludes that while these tools offer alternative protection routes, they risk diluting the sui generis status of GIs if not managed with a clear understanding of their distinct legal functions.

References

Audier J (2003) « Quelle stratégie juridique pour la commercialisation du produit agricole? » (Which legal strategy for marketing the agricultural product?) Marques et indications géographiques de la filière vitivinicole », Revue de Droit Rural, n° 311 mars

Baylos Corroza H (1993) Tratado de Derecho Industrial (Handbook on Industrial Property), 2ª ed, Civitas

Bercovitz Rodriguez-Cano A (2000) "Nociones introductorias sobre denominaciones de origen e indicaciones geográficas protegidas" (Preliminary notions on designations of origin and geographical indications of origin) in Homenaje a M. Currell, Estudios sobre Propiedad Industrial, Grupo Español de la AIIPI; Barcelona, p. 85 y ss

Botana Agra M (2001) Las Denominaciones de Origen, Tratado de Derecho Mercantil (The designations of origin. Handbook of Commercial Law), tomo XX, vol 2°, Marcial Pons, Madrid

De la Calle Robles L (2002) "Denominaciones de Origen y protección económica", (Designations of origin and economic protection) Estudios Agrosociales y Pesqueros, MAPA, n° 194, p.s. 27–48

De Martín Muñoz A (2003) "Marcas colectivas y denominaciones geográficas en la Ley 17/2001, de Marcas" (Collective trade marks and geographical indications in Spanish Trade Mark Act 17/2001), Revista de Derecho Mercantil, n° 247, enero-marzo

Gevers J (1990) 'Geographical Names and Signs Used as Trade Marks', European Intellectual Property Review, 8:285–91

Grazioli A (2003) « Les principes « d'exclusivité de temps et de territorialité » Quel sens dans le contexte de conflits entre marques et indications géographiques? » (The principles of exclusivity and territoriality: which role in the conflicts between trade marks and geographical indications), AIDV Bulletin n° 34, pp 35–44

Guillem Carrau J (2021) Denominaciones geográficas de calidad. Estudio de su régimen jurídico en la OMC, la Unión europea y el Derecho español (Quality Geographical Indications. A study of the legal schem of the WTO, EU and Spanish law), Tirant lo blanch, 2 ed.

Guillem Carrau J (2024) An insight into the EUIPO Board of Appeal approach to GIs EU law. E.I.P.R. 46(12):749–765

Günzel, O. (2012) 'Opposing Third Parties' Applications for Geographical Indications under Regulation 510/2006', Journal of Intellectual Property Law & Practice 7(4): 244–250

Knaak R (2015) Geographical indications and their relationship with trade marks in EU law. IIC Int Rev Intellect Prop Compet Law 46:843–867

Largo Gil R (2006) Las marcas colectivas y las marcas de garantía (Collective and certification trade marks), 2 ed., Madrid, Thomson Civitas

Leach Ros B (2002) "Las denominaciones de origen agroalimentarias" (The designation of origin in the food sector). Revista Aragonesa de Administración Pública, n °21, pp 321–382

López Benítez M (1996) Las Denominaciones de Origen (The designations of origin). CEDECS, Barcelona

Marchisio E (2020) Are EUIPO trade mark examiners and the board of appeal bound by precedent decisions? GRUR Int 69(9):893–901

Maroño Gargallo MM (2002) La protección jurídica de las denominaciónes de origen en los derechos español y comunitario (The legal protection of the designation of origin in Spanish and EU law). Marcial Pons, Madrid

Medina González A (2012) Protection of geographical indications against translation, generic use, evocation, and other potential enemies. J Intellect Property Law Pract 7(1):20–22

Montero García-Noblejas P (2023) La marca de certificación de la Unión Europea (The EU certification trade mark). Tirant lo Blanch, Valencia

O'Connor B (2004) The law of geographical indications. Cameron May, London.

Pérez Ferreras S (2003) « Quelles sont les différences juridiques qui permetent la distinction entre les indications géographiques et les marques? » (Which are the differences that allow the distinction among GIs and trade marks). AIDV Bulletin n°34, pp 14–20

Reichert D (2003) Are conflict rules for geographic indications or trade marks (such as "First in time") applicable in conflicts between these two types of intellectual property rights or not? AIDV Bulletin n° 34, pp 20–23

Stern S (2003) "Geographical indications and trade marks: conflicts and possible resolutions" en Worldwide Symposium on Geographical Indications, WIPO, 2003 y en AIDV Bulletin n° 34, pp 44–50

Zafrilla Díaz-Marta V, Kyrylenko A (2021) The ever-growing scope of Geographical Indications' evocation: from Gorgonzola to Morbier. J Intellect Property Law Pract 16(4–5):442–449

Conclusions

8

Abstract

The EU's Geographical Indications (GIs) scheme stands as a pivotal instrument within the Common Agricultural Policy and the internal market, balancing economic, cultural, and legal interests while navigating persistent debates. This analysis demonstrates how landmark judgments—from *Feta* and *Prosciutto* to *Manchego* and *Nero Champagne*—have shaped the scope, enforcement, and procedural integrity of GI protection, reinforcing its role in intellectual property, trade policy, and sustainability. The interplay between registration requirements, Member State obligations, and EU oversight reveals both the strengths and systemic challenges of the scheme, particularly in addressing infringements, genericity risks, and digital enforcement under the Digital Services Act. By examining the tensions between GIs and trade marks, the limits of labelling rules, and the prohibition of evocation and misuse, this work underscores the evolving jurisprudence that defines the generous protection of the GI under EU Law. Ultimately, the EU's GI system reflects a delicate equilibrium—one that safeguards terroir, savoir-faire, and economic value while adapting to global trade dynamics and legal innovation. For professionals engaged in agricultural law, intellectual property, or international commerce, these insights offer a critical foundation for understanding the past, present, and future trajectory of GI protection in the EU.

This book has provided a comprehensive review of the landmark EU cases on GIs, inviting readers to reexamine the nature and role of GIs and their fair nature of being intellectual and commercial property rights in the framework of the internal market of the European Union and their interaction and coexistence with the trade marks. Moreover, gaining insight into the judgments related to GIs is essential to understanding how the complex puzzle of GI protection fits together within EU law.

© The Author(s), under exclusive license to Springer Nature Switzerland AG 2026

J. Guillem Carrau, *Landmark EU Geographical Indications Case Law*, Law for Professionals, https://doi.org/10.1007/978-3-032-26101-4_8

As in any other field of the *acquis communautaire*, the judgments of the European Courts ensure the uniform application and interpretation of EU law in the field of GIs. However, EU law on GIs is not an easy sailing subject. For many years, the CJ and the GC have been weaving and embroidering a pattern that could help to understand how the protection of this specific intellectual property right is defined and its own evolution. Furthermore, the landmark case law on the subject matter always contribute to a public thought-provoking debate on the GI's protection.

A review of the EU case law reveals a clear trend in favor of protecting GIs, reinforcing the importance of this category of intellectual property right. The leading cases of *Feta*, *Prosciutto* and *Mozzarella* have paved the way for a clearer understanding of the EU GIs system, ultimately revealing its primary objective as a part of the Common Agricultural Policy. Through a series of notable cases, including *Rioja II* and *Schwarzwälder Schinken*, the GIs have progressively defined the boundaries of the Internal Market, uncovering the complex relationships between GIs, intellectual property, and trade. In the crossroads of the common agricultural policy and the internal market, GIs also are a crucial element of the common commercial policy as demonstrated in *Commission v Council (Revised Lisbon Agreement)* and their presence in international markets inform the negotiations of bilateral and multilateral trade agreements by the EU.

In fact, let's highlight that, in the EU, GIs are defined as an economic driver in the common agriculture policy, a competitiveness enhancer in the EU's internal market and a strategic asset and a global influence multiplier in the common commercial policy.

We are confronted to a GIs' unitary and exhaustive system focussed on a set of key elements. *Rioja II* and *Turrón de Jijona* help to clarify that a GI constitutes an intellectual and commercial property right and protects the origin and its characteristics as related in *Sekt and Weinbrand* and somehow the intrinsic reputation as discussed in *Nero Champagne*. The legislative position and historical development of a GI as a key element to protect it have been highlighted in cases such as *Grana Biraghi,* while *Warstein* has helped identify what is exactly protected under the GI system.

The system relies on the existence of an EU title obtained through a registry, as described in *Salame Felino*, and prevents Member States from running separate national or regional schemes (*Budějovický Budvar beer*). The scheme is only available to eligible producers who respect the product specification, as declared in *Manchego*. Finally, the EU legislator emphasizes the concept of sustainability as a key component for the future and, based on the existing case law, has done an update of the EU GIs scheme by adopting Regulation (EU) 2024/1143.

The unity of the system stems from the fact that, as said in *Porto Charlotte*, a registration is *conditio sine qua non* for the protection and it has constitutive character. Procedural aspects, as the role of the different administrations through a division of tasks for registration, controls and enforcement, have been delimitated in cases such as *Halloumi* cheese and *La Conqueste, Kühne* and *Jambon sec de Corse*. Other cases, like *Comté* and *Piadina*, have dispelled what could happen when national decisions are reviewed by national courts. We reached logical conclusions on the troubles faced when the national decision is reviewed by a national Court or when the Member State is called to adopt measures against conduct which infringe a PDO.

Due to the adoption of the EU Digital Services Act, new obligations emerged for national authorities as regards enforcement and online platforms and providers of hosting services. The respect and protection of GIs in online markets depend on the development and implementation of these measures that will help the producers to react against online illegal content. The relevant national judicial or administrative authorities will be able to issue an order to act against such illegal content.

We have described the genericity thread as the Elephant in the GIs Room. It shows that GIs' protection does not encompass generic terms or elements, as consistent with relevant case law, like *Aceto Balsamico di Modena*.

In the GIs field, the EU Courts run a case-specific race to the bottom which have resulted on an expanded scope of protection of GIs against evocation. The singular extension of GIs protection as intellectual property right is described, as regards situations of direct or indirect use or misuse in cases like *Bavaria, Chianti, Cuvée Palomar, Riojavina, Colombiano Coffee House.*

The concept of evocation has been clarified, establishing that it does not require the product protected by a GI and the product or service covered by the disputed sign to be identical or similar. Instead, evocation is established when the use of a name creates a sufficiently clear and direct link between that name and the GI in the mind of an average European consumer.

Key Judgments like *Champanillo, Manchego*, and *Morbier* have further defined the concept of evocation. 'Evocation' shield remains the greatest asset for GI holders, often providing a wider scope of protection than traditional 'likelihood of confusion' metrics. Notably, the protection of GIs extends beyond mere conceptual evocation, also encompassing cases involving figurative signs and even product shapes; while *Scotch Whisky Association* has helped understand when a GI faces practices liable to mislead consumers about the true origin of a product. *Salami Felino* has highlighted the application of GI protection independently of national unfair competition rules.

Champagner sorbet and *Weingut* have addressed the use of GIs as ingredients and labelling and presentation in the wine sector. GIs labelling rules are an expression of the exclusive use of the GI and they are interpreted in connection to labelling rules. The GIs labelling rules underpin the exclusive use of GIs including the compulsory use of Union symbols within the indications contained in the Commission's Guidelines on labelling practices. However, the examination must be done equally as regards the Food Labelling Regulation.

An insight into the Decisions of the EUIPO Boards of Appeal on GIs helps to understand how the protection of the GIs as a specific intellectual property right weight gainer in many litigations because of the singular position is, is assigned to GIs by the European legislator in the IP field. The examination of the decisions of the Boards of Appeal illustrates that, when coexisting with trade marks, the EU legislator recognises a predominant position to GIs. A name must not be registered as a GI where, in the light of a trade mark's reputation and renown and the length of time that it has been used, registration of the name proposed as a GI would be liable to mislead the consumer as to the true identity of the product. Conversely, a trade mark consisting of a GI, or that contains a GI, that could mislead the consumer about the true origin of the product, should not be protected. The analysis of the referred decisions serves to draw conclusions on the practice and interpretation of Article 7(1)(j) EUTMR which provides a specific ground for refusal to EUTMs examiners, using the *BNIC* criteria to assess the existence of evocation as an absolute ground for refusal.

First and foremost, a review of the recent Boards of Appeal's decisions on GI's permitted musing the concept of comparable goods, the restrictions of goods and services and when not to look for a EUTM with a GIs reference. It is also part and parcel of our effort in this book to pay attention to GIs protection behind the inter partes proceedings' scenes by scrutinizing decisions on the GIs scope of protection against direct and indirect use and reputation, misuse, evocation, and misleading indications. On top of that, there are some Board of Appeal Decisions on procedural aspects that will allow to identify some particularities in appeal proceedings related to GIs like a request for GIs external expertise, the substantiation of the GIs' right or the right to be heard.

To sum up, the Court of Justice noted, in *Nero Champagne*, that EU law does not prohibit a trade mark from containing a GI. However, its registration may be refused if the PDO does not comply with the product specification concerned, if its use exploits the reputation of a PDO or if the mark applied for conveys a false or misleading indication as to the provenance or origin of the product.

The EU legislator considers that it is possible to register geographical indications as EU collective trade marks. The review of the *Halloumi* and *Darjeeling tea* Judgments offers an opportunity to reflect on the weakening of a GI when doing so. The approach concludes with an attempt to explain why GIs are not classified under certification schemes or certification trademarks at EU level.

As said in the introduction of the book, this system empowers producers to manage their respective GI, enabling them to respond to growing societal demands for sustainable products. The scope of protection is carefully calibrated to balance the intellectual property rights of agrifood GIs with the coexistence of pre-existing trademarks. Notably, the protection of GIs extends beyond mere conceptual evocation, also encompassing cases involving figurative signs and even product shapes. A review of European Court of Justice and General Court case law reveals a clear trend in favour of protecting GIs, reinforcing the importance of this intellectual property right. GIs protection is noteworthy politically desired.

All in all, one of the key challenges of the EU's GIs scheme is striking a balance between protecting the intellectual property rights of producers and ensuring that the system is fair to all.

Annex: Selected Geographical Indications Case Law

Court of Justice of the European Union

T-114/99, CSR PAMPRYL v Commission, ECLI:EU:T:1999:281
T-215/00, La Conqueste v Commission, ECLI:EU:T:2001:23
T-291/03, Consorzio per la tutela del formaggio Grana Padano v OHMI - Biraghi (GRANA BIRAGHI), EU:T:2007:255
T-53/04 to T-56/04, T-58/04 and T-59/04, Budějovický Budvar v OHMI - Anheuser-Busch (BUDWEISER) ECLI:EU:T:2007:167
T-60/04 to T-64/04, Budějovický Budvar v OHMI - Anheuser-Busch (BUD), ECLI:EU:T:2007:169
T-57/04 and T-71/04, Budějovický Budvar v OHMI - Anheuser-Busch (BUDWEISER), ECLI:EU:T:2007:168
T-318/06 to T-321/06, Moreira da Fonseca/OHIM — General Óptica (GENERAL OPTICA), EU:T:2009:77
T-485/07, olive Line International v OHIM — Knopf (Olive), not published, EU:T:2011:467
T-345/08 and T-357/08, Botolist/Botocyl, EU:T:2010:529
T-237/08, Abadía Retuerta v OHIM (CUVÉE PALOMAR), EU:T:2010:185
T-138/09, RIOJAVINA, EU:T:2010:226
T-534/10, HELLIM / HALLOUMI, EU:T:2012:292. And C-393/12P, HELLIM / HALLOUMI, EU:C:2013:207
T-143/11, Consorzio vino Chianti Classico v OHMI - FFR (F.F.R.), EU:T:2012:645
T-359/14, Federación Nacional de Cafeteros de Colombia/OHIM — Accelerate (COLOMBIANO COFFEE HOUSE), not published, EU:T:2015:651
T-659/14, Instituto dos Vinhos do Douro e do Porto v OHMI - Bruichladdich Distillery (PORT CHARLOTTE) ECLI:EU:T:2015:863
T 43/15, *CRM* v *Commission* not published, EU:T:2018:208
T-41/16, Cyprus Turkish Chamber of Industry and Others v Commission, ECLI:EU:T:2016:613
T-828/16, Consejo Regulador de la Denominación de Origen "Torta del Casar" v European Union Intellectual Property Office, ECLI:EU:T:2017:918
T-774/16, Cave de Tain, ECLI:EU:T:2018:441

© The Editor(s) (if applicable) and The Author(s), under exclusive license to Springer Nature Switzerland AG 2026

J. Guillem Carrau, *Landmark EU Geographical Indications Case Law*, Law for Professionals, https://doi.org/10.1007/978-3-032-26101-4

T-34/22, Cunsorziu di i Salamaghji Corsi – Consortium des Charcutiers Corses and Others v Commission, EU:T:2023:386

T-23/24, IVDP v EUIPO - Vinoquel- Vinhos Oscar Quevedo (Quevedo Port), ECLI:EU:T:2025:182 T-40/24 - Instituto dos vinhos do Douro e do Porto v EUIPO - The Benriach Distillery Company (PORTSOY) ECLI:EU:T:2025:183

T-425/23, Consejo Regulador «Aceite de Jaén» v EUIPO - Agrícola La Loma (VEGA DEL OBISPO BIO Jaén PRODUCTOS ECOLÓGICOS), ECLI:EU:T:2024:768

T-239/23, Comité interprofessionnel du vin de Champagne and INAO v EUIPO - Nero Lifestyle (NERO CHAMPAGNE), ECLI:EU:T:2025:638

General Court of the European Union

C-181/73, Haegemann v Belgian State, EU:C:1974:41

C-12/74, Commission v Germany, ECLI:EU:C:1975:23

C-193/80, Commission v Italy ECLI:EU:C:1981:298

C-178/84, Commission v Germany. ECLI:EU:C:1987:126

C-182/84, Miro. ECLI:EU:C:1985:470

C-407/85, 3 Glocken and Others v USL Centro-Sud and Others, ECLI:EU:C:1988:401

C-12/86, Demirel v Stadt Schwäbisch Gmünd, EU:C:1987:400

C-286/86, Ministère public v Deserbais ECLI:EU:C:1988:434

C-298/87, Smanor. ECLI:EU:C:1988:415

C-210/89, Commission v Italy, ECLI:EU:C:1990:357

C-269/89, Bonfait, ECLI:EU:C:1990:399

C-47/90, Delhaize Frères v Promalvin and Others, ECLI:EU:C:1992:250

C-3/91, Exportur, ECLI:EU:C:1992:420

C-87/97, Cambozola, EU:C:1999:115

C-321/94 to C-324/94, *Pistre and Others* ECLI:EU:C:1997:229

C-388/95, Belgium v. Spain, ECLI:EU:C:2000:244

C-108/97 and C-109/97, Chiemsee, EU:C:1999:230

C-129/97 and 130/97, Chiciak and Fol ECLI:EU:C:1998:274

C-312/98, Haus Cramer, ECLI:EU:C:2000:599

C-447/98 P,-Molkerei Großbraunshain and Bene Nahrungsmittel v Commission, ECLI:EU:C:2000:586

C-269/99, Carl Kühne, ECLI:EU:C:2001:659

C-66/00, Bigi ECLI:EU:C:2002:397

C-469/00, Ravil, ECLI:EU:C:2003:295

C-108/01, Consorzio del Prosciutto di Parma and Salumificio S. Rita, EU:C:2003:296

C-465/02 and C-466/02, Denmark v Commission, Germany v Commission, ECLI:EU:C:2005:804 and ECLI:EU:C:2005:636

C-347/03, Regione autonoma Friuli-Venezia Giulia and Agenzia regionale per lo sviluppo rurale (ERSA), EU:C:2005:285

C-132/05, Commission v Germany, EU:C:2008:117

C-343/07, Bayerisches Bier ECLI:EU:C:2009:415

C-446/07, Alberto Severi v Regione Emilia Romagna ECLI:EU:C:2009:530
C-4/10 and C-27/10, *BNI Cognac*, EU:C:2011:484
C-35/13, Assica and Krafts Foods Italia, ECLI:EU:C:2014:306
C-517/14 P, Schutzgemeinschaft Milch und Milcherzeugnisse v Commission, EU:C:2015:700
C-75/15, Viiniverla EU:C:2016:35
C-675/15 P, (Joined Cases C-673/15 P, C-674/15 P, C-675/15 P, C-676/15 P) The Tea Board v EUIPO, ECLI:EU:C:2017:702
C 56/16 P, PORT CHARLOTTE, EU:C:2017:693
C-393/16, Comité Interprofessionnel du Vin de Champagne, ECLI:EU:C:2017:991
C-454/16 P to C-456/16 P and C-458/16 P, Global Steel Wire and Others v Commission, not published, EU:C:2017:818
C-44/17, Scotch Whisky Association, ECLI:EU:C:2018:415
C-367/17, Schwarzwälder Schinken, ECLI:EU:C:2018:1025
C-614/17, Fundación Consejo Regulador de la Denominación de Origen Protegida Queso Manchego, EU:C:2019:344
C-432/18, Consorzio Tutela Aceto Balsamico di Modena. ECLI:EU:C:2019:1045
C-569/18, Caseificio Cirigliana and Others, ECLI:EU:C:2019:873
C-785/18, GAEC Jeanningros, ECLI:EU:C:2020:46
C 490/19, Syndicat interprofessionnel de défense du fromage Morbier, ECLI:EU:C:2020:1043
C-783/19, Comité Interprofessionnel du Vin de Champagne v GB ECLI:EU:C:2021:350
C-159/20, Commission v Denmark (AOP Feta), ECLI:EU:C:2022:561
C-354/22, Weingut A, ECLI:EU:C:2023:916
C-579/23 P, Consortium des Charcutiers Corses ECLI:EU:C:2024:832
T-239/23, Comité interprofessionnel du vin de Champagne and INAO v EUIPO - Nero Lifestyle (NERO CHAMPAGNE)

EUIPO Boards of Appeal

26/10/2012, R 1732/2012, *Cazorliva*
16/01/2013, R 659/2012-5, *Fratelli Parmigiani Alta Salumeria Salame di Parma*
02/12/2013, R 1299/2013-1, Vana Tallinn Authentic Estonian Liqueur
09/01/2014, R 426/2013-1, *Azienda Olearia Del Chianti*
16/01/2014, R 51/2013-1, La Croix Des Papes
20/01/2014, R 1900/2013-5, *Parmatutto*
27/03/2014, R 1200/2013-5, *Colombiano Coffee House*
18/11/2014, R 2462/2013-2 – TOROLOCO (FIG. MARK)
23/02/2016, R 612/2015-4, *Toro*
01/06/2016, R 2531/2015-2, *Ibiza Flirt*
15/06/2016, R-1105/2015-4, *Portobello Road No 171*
04/04/2017, R 1428/2016-2, ORGANIC Quijote (fig.) / m LA MANCHA DENOMINACIÓN DE ORIGEN (fig.) et al.
15/09/2017, R 500/2017-2, CAVA DE PARAJE CALIFICADO

29/09/2017, R 1006/2017-5, CA' DEL MAGRO, § 30 (CA' DEL MAGRO / VAL DI MAGRA
18/01/2018, R 1229/2017-5, OPTIMA feinkost Olympia (fig.)
25/01/2018, R 674/2017-1, YORKSHIRE PROVENDER
05/02/2018, R 1108/2017-2, Matusalem Espíritu de Cuba/Cuva
21/02/2018, R 2305/2017-5, CRU CÔTES DU RHÔNE VACQUEYRAS (fig.)
23/03/2018, R 26/2018-5, MATPRAT (fig.)
21/06/2018, R 2079/2017-1, CASA FONDATA NEL 1991 FRATELLI VINO BIANCO FRATELLI Asti (fig.) / CONSORZIO DELL'ASTI D.O.C.G. (fig.) et al.
16/07/2018, R 2110/2017-1, PIEMONTINO (fig.)
30/11/2018, R 251/2016-1, COLOMBIANO COFFEE HOUSE (fig.)
07/12/2018, R 477/2017-1, CAFE Gran Colombiano con todo el aroma y sabor del buen café (fig.)
07/03/2019, R 831/2018-2, PUERTO FINO (fig.), § 28-31 (PUERTO FINO / PORTO)
23/04/2019, R 2723/2017-5, Lucky Lasso Kentucky Straight Bourbon Whiskey / Protected designation of origin or a geographical indication Bourbon Whiskey
02/05/2019, C-614/17, Queso Manchego, EU:C:2019:344, § 50 (ROCINANTE / QUESO MANCHEGO)
03/07/2019, R 942/2018-4, BEHER Bernardo Hernandez BEHER GUIJUELO (fig.) / Guijuelo Denominación de Origen (fig.) et al.
19/07/2019, R 2191/2018-2, Grana padano
19/09/2019, R 1223/2019-5, Mappa di Imola
14/11/2019, R 425/2019-1, By Latúe PRO & ECO (fig.) / PROSECCO PDO PROSECCO DOC (fig.) et al.
20/01/2020, R 1315/2019-5, CASA TIMIŞ (fig.)
17/04/2020, R 1132/2019-4, Champagnola / Champagne
21/04/2020, R 993/2019-2, Portwo gin / Porto
08/04/2020, R 1920/2019-1, Mehr allgäu passt in keinen Käse
11/05/2020, R 2028/2019-2, Port ruighe / PORT/PORTO
13/05/2020, R 2230/2017-2, La farra
16/06/2020, R 2650/2019-5, Château de Pommard
12/08/2020, R 468/2020-5, The Great Earl Irish Whiskey
03/09/2020, R 2751/2019-4, TORO DE PIEDRA (fig.) / Toro denominación de origen et. Al.
15/09/2020, R 2931/2019-1, Eldstickan / SOLSTICKAN (fig.) et al.
03/09/2020, R 2751/2019-4, TORO DE PIEDRA (fig.) / Toro denominación de origen et. Al
02/10/2020, R 282/2020-5, Campo Verde / Vinho verde et al.
20/11/2020, R 1450/2020-5, FENDENTE MAREMMA TOSCANA Le uve vengono raccolte a mano BURATTINI (fig.)
09/02/2021, R 237/2020-4, Sol de Mallorca / Mallorca
18/02/2021, R 1720/2020-5, Delicious Green Eva Fricke (fig.) / Vinho verde et al.
25/02/2021, R 696/2018-5, QUESO Y TORTA DE LA SERENA (fig.) / Torta del Casar et al

04/03/2021, R 148/2021-5, allevatori bufalini campani abc caseificio dal' 78 (fig.)
06/04/2021, R 885/2020-1, Es Valentía / VALENCIA DENOMINACION DE ORIGEN (fig.) et al.
07/04/2021, R 1661/2020-5, GREEN CHOICE (fig.) / Vinho verde et al
07/06/2021, R 1249/2016-1, VALSECCO / VALDOBBIADENE PROSECCO et al.
16/07/2021, R 879/2020-5, Padron AUTÉNTICO (fig.) / Pemento de Herbón
01/09/2021, R 1734/2020-2, Las villas
01/10/2021, R 566/2015-2, COLOMBUENO / CAFÉ DE COLOMBIA
05/11/2021, R 1103/2019-1, PERISECCO (fig.) / PROSECCO (other) et al.
22/04/2022, R 981/2021-1, CHENINE Brut PREMIUM CAVA BOTTLE (3D) / CONSEJO REGULADOR CAVA (fig.) et al.
08/03/2022, R 803/2021-2, Teslaquila / TEQUILA (GI)
21/03/2022, R 2564/2019-2, BOLGARÉ (fig.) / Bolgheri et al.
04/04/2022, R 101/2022-5, ROMITORO / Toro
11/04/2022, R 100/2022-5, ROMITÒRO (fig.) / Toro
20/04/2022, R 1296/2021-4, Romeral de jabugo / Jabugo
22/04/2022, R 981/2021-1, CHENINE Brut PREMIUM CAVA BOTTLE (3D)/ Council REGULADOR CAVA (fig.) et al.
17/02/2023, R 531/2022-2, NERO CHAMPAGNE / Champagne (GI)
12/05/2022, R 1941/2021-5, HALLOUMAKI (fig.) / Halloumi
25/07/2022,R1939/2021-5,Halime(fig.)/'ΧΑΛΛΟΥΜΙ'(HALLOUMI)/'HELLIM'
29/07/2022, R 1940/2021-5, Macroom buffaloumi / Halloumi
29/08/2022, R 462/2022-5, GRILLOUMAKI / Χαλλουμι' (Halloumi)/'Hellim
01/09/2022, R 1714/2021-5, THE GRAND WINES GW RIOJA ALAVESA (fig.) / Rioja (DENOMINACION DE ORIGEN PROTEGIDA) et al
26/11/2022, R 2242/2020-1, MORELLONE LE CANIETTE (fig.) / Morellino et al.
29/03/2023, R 1667/2021-4, MAISON SICILE (fig.)
01/06/2023, R 1119/2022-1, VEGA DEL obispo BIO Jaén organic PRODUCTOS (fig.)
15/06/2023, R 1761/2022 - 5, MARQUÉS DE LA MANCHA / LA MANCHA DENOMINACION DE ORIGEN (fig.) et al.
07/08/2023, R 2248/2022-4, AMARTE TEQUILA 100% AGAVE (fig.) / TEQUILA (GI)
09/08/2023, R 2272/2022-4, ADICCION TEQUILA (fig.) / TEQUILA
13/10/2023, R 825/2023-5, CAVCA / Café de Cauca (GI)
07/11/2023,R937/2022-4,CONDIPARMA(fig.)/PARMIGIANOREGGIANO(GI)
14/11/2023, R 2471/2022-4, Quevedo Port / Port
14/11/2023, R 1885/2022-4, PORTSOY / Port
14/11/2023, R 37/2023-4, PORTUS SANTA MARIA / Port
15/11/2023, R 1073/2022-5, GRANA PADANO (fig)
05/12/2023, R 1120/2023-2, FÉLIX de Múrtiga JABUGO (fig.) / FELIX et al.
20/02/2024, R 1454/2022-5, PriSecco / 'Prosecco'
06/03/2024, R 1033/2023-2, DUTCH GENQUILA / TEQUILA (GI)

GPSR Compliance
The European Union's (EU) General Product Safety Regulation (GPSR) is a set of rules that requires consumer products to be safe and our obligations to ensure this.

If you have any concerns about our products, you can contact us on

ProductSafety@springernature.com

In case Publisher is established outside the EU, the EU authorized representative is:

Springer Nature Customer Service Center GmbH
Europaplatz 3
69115 Heidelberg, Germany

www.ingramcontent.com/pod-product-compliance
Ingram Content Group UK Ltd.
Pitfield, Milton Keynes, MK11 3LW, UK
UKHW020746021026
476198UK00003B/80
* 9 7 8 3 0 3 2 2 6 1 0 0 7 *